'H' Grade Chemistry Essential Facts and Theory

E.R.Allan and J.H.Harris

Edward Arnold

First published 1978
by Edward Arnold (Publishers) Ltd
41 Bedford Square, London WC1B 3DQ

Edward Arnold (Australia) Pty Ltd
80 Waverley Road, Caulfield East
Victoria 3145, Australia

Repinted 1980, 1985

Printed and bound at The Bath Press, Avon

Contents

Preface

Our aim in this book has been primarily to produce a concise summary of the final year's work for the Scottish Certificate of Education Higher Grade Examination in Chemistry. However, since some topics are found by pupils to be considerably more difficult than others, we felt justified in elaborating to some extent in certain chapters such as those dealing with the mole, electrode potentials, thermochemistry and equilibrium. This text assumes a knowledge of Chemistry up to S.C.E. 'O' Grade, but where topics occur at both levels then the essential background is also covered in this book.

So far as is possible, practical details of experiments have only been included where they are considered to be essential to a clear understanding of the factual material. This is not to deny the fundamental importance of practical work in any Chemistry course and we recognise the debt owed to pioneering authors at the time of the emergence of the 'Alternative Syllabus' as it was then known. However, we believe that teachers should be free to include the experiments that they prefer. Throughout the text, reference is made to experimental results and conclusions, frequently illustrated with diagrams, graphs or tables.

A frequent complaint of many teachers is the paucity of suitable examples for practice. Consequently, we have included worked examples in those chapters which deal with numerical topics, such as the mole and thermochemistry, and we have appended several questions to appropriate chapters.

E.R.A.
J.H.H.

Acknowledgements

The authors wish to acknowledge the many valuable suggestions for improvements from our colleagues and to express our thanks to Mrs. M. Harris for preparing a substantial part of the original typescript.

The Publishers wish to thank the Scottish Certificate of Education Examination Board for permission to reproduce those examples indicated in the text by S.C.E.E.B. and also Blackie & Son Ltd. for permission to reproduce the tables which appear at the end of the book. They would also like to thank B.P. for permission to reproduce the photograph used on the cover.

1 Nuclear Chemistry

Atomic Structure

Atoms consist of a nucleus and extranuclear electrons. The nucleus is made up of neutrons and protons.

Table 1.1

	mass (approximate)	charge	location	symbol
proton	1	+1	nucleus	$^{1}_{1}p$ or $^{1}_{1}H$
neutron	1	0	nucleus	$^{1}_{0}n$
electron	1/2000	−1	orbitals	$^{0}_{-1}e$

Atomic Number = number of protons (=number of electrons, in an uncharged atom)
Mass Number = number of protons + number of neutrons

Isotopes

These are different types of atoms of the same element i.e. they have the same atomic number but different mass numbers, owing to different numbers of neutrons. Isotopes cannot be distinguished by chemical means, since chemical properties depend on electron arrangement.

We represent a particular isotope of an element as, for example, $^{238}_{92}U$ i.e. the Uranium (atomic no.92) isotope of mass number 238.

Previously we have been concerned solely with the extranuclear electrons, however, a small part of our work is concerned with the effects of changes in the nucleus.

In 1896, Becquerel found that compounds of uranium emit an invisible radiation which will penetrate opaque materials and fog photographic plates. The phenomenon became known as **radioactivity.** Compounds of thorium behave in a similar fashion, as do natural isotopes of some other elements and artificial isotopes of most elements. The radiation was found to be affected by a magnetic field or an electrical field.

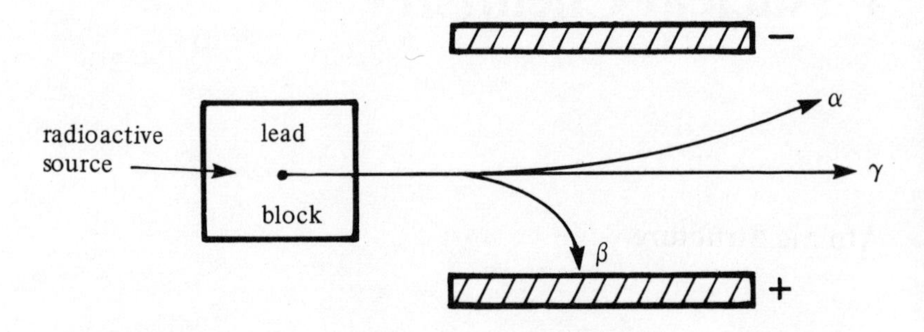

Fig. 1.1. Radiation deflected by an electric field between charged plates.

There are three types of radiation. Other experiments provide the following information:

Table 1.2

name	penetration	nature	symbol	charge	mass
α (alpha)	few cm in air	He nucleus	$^{4}_{2}He$	2+	4
β (beta)	thin metal foil	electron	$^{0}_{-1}e$	1−	1/2000 (approx.
γ (gamma)	great thickness of concrete	E.M.R.	none	none	none

γ radiation is non-particulate. It is electro-magnetic radiation, similar to, but of higher energy than X-radiation. It is emitted alone or accompanying α or β radiation. Radiation is independent of the physical or chemical state of the element (i.e. independent of electronic arrangement) which indicates that it is connected solely with the nucleus. Since radiation originates in the nucleus, the loss of α and β particles changes it considerably.

Loss of an α particle

This is equivalent to losing 2 protons (**decreasing** the atomic number by 2) and 2 neutrons (overall decrease of mass number by 4). For example:

$$^{232}_{90}Th \longrightarrow {}^{228}_{88}Ra + {}^{4}_{2}He$$

The total mass number (superscript) must be the same on each side of the equation, as must be the total atomic number (subscript).

Note. Since the atomic number of the product is 88 it is now radium. Similarly:

$$^{220}_{86}Rn \longrightarrow ^{216}_{84}Po + ^{4}_{2}He$$

Loss of β particle

The β particle is an electron. Since the nucleus does not contain electrons, it is believed to be formed by

$$\underset{\text{neutron}}{^{1}_{0}n} \longrightarrow \underset{\text{proton}}{^{1}_{1}p} + \underset{\text{electron}}{^{0}_{-1}e}$$

Hence loss of the electron results in a gain of one unit of atomic number (one proton) without any change in mass number, since the mass of the proton is almost the same as the mass of the neutron. For example:

$$^{228}_{88}Ra \longrightarrow ^{228}_{89}Ac + ^{0}_{-1}e$$

Again, the total atomic number and mass number must be the same on each side of the equation. Since the product has atomic number 89 it is now actinium. Similarly:

$$^{216}_{84}Po \longrightarrow ^{216}_{85}At + ^{0}_{-1}e$$

In fact the various 'daughter nuclei' are usually radioactive themselves, so that a whole series of radioactive disintegrations occurs until a stable isotope is reached – frequently an isotope of lead.

Artificial radioactivity

Many artificially radioactive isotopes are known. These can be made by bombarding stable isotopes with neutrons in a nuclear reactor. For example:

$$^{27}_{13}Al + ^{1}_{0}n \longrightarrow ^{24}_{11}Na + ^{4}_{2}He$$

This nuclear reaction is sometimes written as $^{27}_{13}Al\ (n, \alpha)\ ^{24}_{11}Na$. The sodium isotope produced then decays by β emission.

$$^{24}_{11}Na \longrightarrow ^{24}_{12}Mg + ^{0}_{-1}e$$

Similarly:

$$^{32}_{16}S + ^{1}_{0}n \longrightarrow ^{32}_{15}P + ^{1}_{1}p \text{ (a proton)}$$

or $^{32}_{16}S\ (n, p)\ ^{32}_{15}P$.

The phosphorus isotope produced is a β emitter:

$$^{32}_{15}P \longrightarrow ^{32}_{16}S + ^{\ 0}_{-1}e$$

The 'target' material for the neutrons is often a compound including the starting isotope. The other part of the compound may also be affected by the neutrons, so selection of the compound is a specialist task.

Half-life

Radioactive decay of the atoms of an isotope is such that the activity of the isotope decreases by half in a fixed time, called the half-life of the isotope. e.g. For an isotope of half-life 2 days the activity is halved in 2 days, becomes ½ of ½ in 4 days and ½ of ¼ in 6 days etc.

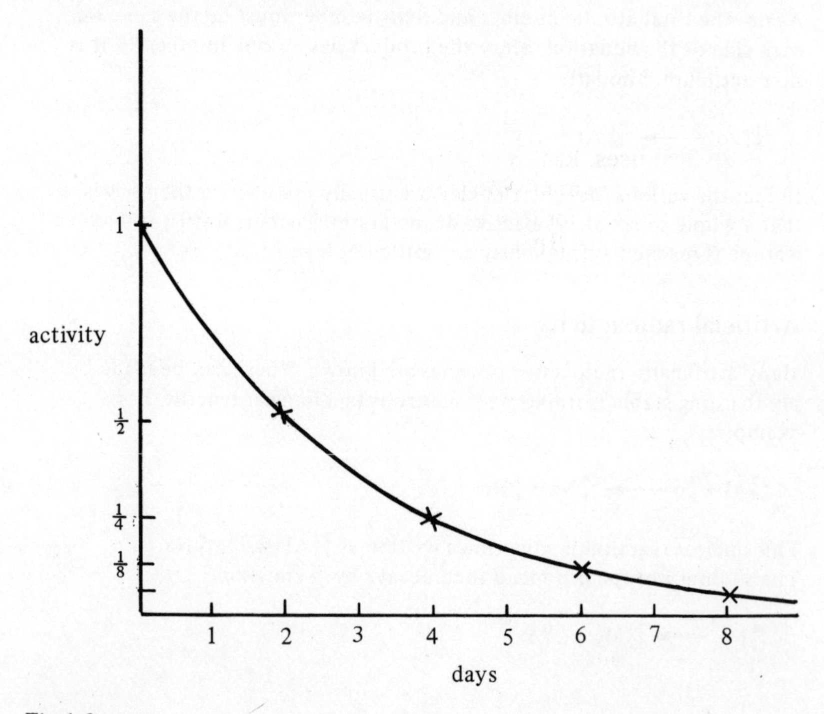

Fig 1.2

Half-lives vary from seconds to millions of years and are characteristic of particular isotopes.

The half-life of any isotope is independent of the mass of the sample being investigated. It is independent of temperature, pressure, concentration, presence of catalysts or chemical state of the isotope. In the half-life, half of the atoms of the isotope decay, but the process is completely random and it is not possible to predict the time of decay of any individual atom. Half-life is often abbreviated to $t_{1/2}$.

The uses of radio isotopes

This is an ever expanding field, and so only certain important representative examples are given.

1. **Industrial radiography.** A source of very penetrating γ radiation ($^{60}_{27}Co$ or $^{192}_{77}Ir$ being commonly used) is used to examine castings and welds for imperfections. A photographic film can be used as the detector, as in X-ray work.

2. **Tracer use.** The uptake of certain elements by plants and animals can be traced by using a radioactive isotope of the particular element, e.g. the uptake of iodine by the thyroid gland can be traced by $^{131}_{53}I$.

3. **Medical uses.** Radioactive isotopes have been used for a considerable time for the treatment of various types of cancer. For example the γ emitter $^{60}_{27}Co$ is used for treatment of deep-seated tumours and the less penetrating β emitter $^{32}_{15}P$ is used for treating skin cancer by direct application to the affected area.

4. **Detecting flow patterns.** Leaks in pipelines, flow patterns in estuaries and ventilation flows can be investigated by using small quantities of short half-life isotopes which leave negligible residual activity.

5. **Dating.** Because radioactive isotopes are constantly decaying at a known rate, the age of materials containing them can be estimated by finding the present content of the isotope. For example $^{14}_{6}C$ is used to date archaeological specimens between about 600 and 10 000 years old. The method depends on the production of $^{14}_{6}C$ in the atmosphere by bombardment of nitrogen with neutrons formed from the effects of cosmic rays on other atoms.

$$^{14}_{7}N + ^{1}_{0}n \longrightarrow ^{14}_{6}C + ^{1}_{1}p$$

This ${}^{14}_{6}C$, like ${}^{12}_{6}C$, is constantly taken up by growing plants and thence by animals, so that their ${}^{14}_{6}C : {}^{12}_{6}C$ ratio is known. When they die however, the ${}^{14}_{6}C$ decays

$$ {}^{14}_{6}C \longrightarrow {}^{14}_{7}N + {}^{0}_{-1}e \ (t_{1/2} = 5600 \text{ years}) $$

By measuring the ${}^{14}_{6}C : {}^{12}_{6}C$ ratio the time since the sample 'died' can be estimated.

Many systems involving long half-life isotopes exist for dating rocks, and for checking the age of vintage wines the short half-life tritium ${}^{3}_{1}H$ has been used ($t_{1/2}$ = 12.4 years).

Example: A sample of wine shows a β emission from tritium ${}^{3}_{1}H$ which is $\frac{1}{8}$ of that from wine bottled at the present day. If the half-life of tritium is 12.4 years, and its concentration in the atmosphere, and in water exposed to the atmosphere, is assumed to have remained constant, for how many years has the wine been bottled?

Activity is $\frac{1}{8}$ of original β activity i.e. $(\frac{1}{2})^3$
i.e. 3 half-lives have elapsed = 3 x 12.4 years
The wine is approximately 37 years old.

Examples for practice

1. Write down the symbol of the particle formed when
   (a) a francium atom loses its outer electron,
   (b) ${}^{223}_{87}Fr$ loses a β particle.

2. Complete the blanks, (a) to (d), in the following radioactive series.

$$ {}^{231}_{90}Th \xrightarrow{-\beta} (a) \xrightarrow{-\alpha} (b) \begin{cases} \xrightarrow{-(c)} {}^{223}_{87}Fr \\ \xrightarrow{-(d)} {}^{227}_{90}Th \end{cases} $$

3. Complete the following equations

(a) ${}^{234}_{90}Th \longrightarrow \quad + {}^{0}_{-1}e$

(b) ${}^{222}_{86}Rn \longrightarrow \quad + {}^{4}_{2}He$

4. Find x, y and z in the following equations

$$ {}^{6}_{3}Li + {}^{1}_{0}n \longrightarrow {}^{3}_{1}H + x $$

$$ {}^{234}_{92}U \longrightarrow y + {}^{4}_{2}He $$

$$ z \longrightarrow {}^{212}_{83}Bi + {}^{0}_{-1}e $$

5. The activity of a sample of $^{131}_{53}I$ is found to be only $\frac{1}{16}$ of the activity when it arrived at a hospital 32 days earlier. What is the half-life of $^{131}_{53}I$?

6. What fraction of the activity of a sample of $^{106}_{45}Rh$ will remain after 3 minutes? ($t_{1/2}$ = 30 seconds).

7. A method of archaeological dating is based on the properties of the isotope carbon-14. Ordinary carbon atoms are nearly all of atomic mass 12, but one atom in every million million has the atomic mass 14. Both isotopes undergo identical chemical reactions, including photosynthesis, respiration, etc., but the atoms of carbon-14 emit low energy β-rays which even when present in very small amounts can be detected by means of a sensitive Geiger-Muller counter.

Carbon-14 atoms are being constantly formed in the upper atmosphere by the reaction of nitrogen with neutrons and continuously decaying into nitrogen, at the rate of half of them disappearing in about 5000 years, so that in the atmosphere and in all living matter there is a fixed proportion of carbon-14. Once the living matter has died, however, it is no longer obtaining fresh supplies of carbon-14 by photosynthesis or by eating plants which have photosynthesised.

The carbon-14 method has given results which agree well with the specimens of known age such as objects found in the Pyramids, but it cannot be used for remains which are more than about 50 000 years old.

(a) Explain why the concentration of radioactive carbon remains constant during the life of the organism.
(b) Indicate clearly the difference in structure of the two isotopes of carbon.
(c) What is the evidence in the passage that carbon-14 is a radio-isotope?
(d) Write an equation to represent the formation of the radio-isotope carbon-14 in the atmosphere.
(e) Write an equation to represent the radioactive decay of carbon-14.
(f) What is the half life of carbon-14?
(g) What fraction of the original radioactivity would remain in a sample of wood 50 000 years old?
(h) Why can this method of dating not be used for remains which are more than 50 000 years old?
(i) What is the evidence in the passage for the alchemists' dream of transmutation?
(j) What does the passage suggest is the major advantage of detection and measurement involving radio-isotopes over normal chemical methods?
(k) The age of clothing found in archaeological 'digs' can be determined by carbon dating; that of diamond jewellery cannot. Give an explanation.

(S.C.E.E.B.)

2 The Mass Spectrometer

Determination of Mass Numbers and Isotope Abundances.

The original function of the mass spectrometer was to determine the mass numbers of the various isotopes of the elements and the proportions of these isotopes in a normal sample of the element. It often provided the first concrete evidence of the existence of isotopes of an element.

The mass spectrometer is basically a means of causing atoms of an element to form positive ions by removing one or more electrons, and then accelerating the ions to pass through a magnetic field of high intensity. Since the beam of ions moving at high speed resembles a current of positive electricity, it will experience a force tending to move it at right angles to the magnetic field and its own direction of travel. It will follow a curved path in the magnetic field. The actual path will depend on the mass and charge of the ion. The effect of a magnetic field on a beam of ions from an element made up of two isotopes is shown in Fig. 2.1.

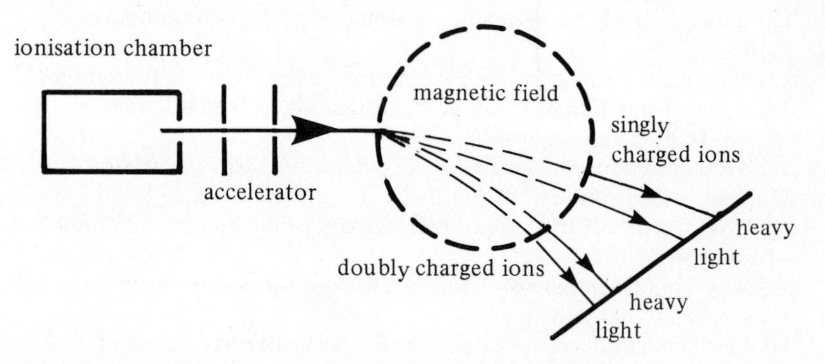

Fig. 2.1

A heavy isotope is deflected to a smaller extent than a lighter one, while a doubly charged ion is deflected more than a singly charged ion. For simplicity we shall only consider ions which have a single charge.

In practice, it is usually more convenient to place the detector at a particular point and vary the magnetic field so that the various beams pass across it in turn. Once the magnetic field control has been calibrated, the mass number of an ion responsible for a particular signal is given directly, and the intensity of the signal is a measure of its proportion of the total. The read-out from the instrument can be given in graphical form as shown below.

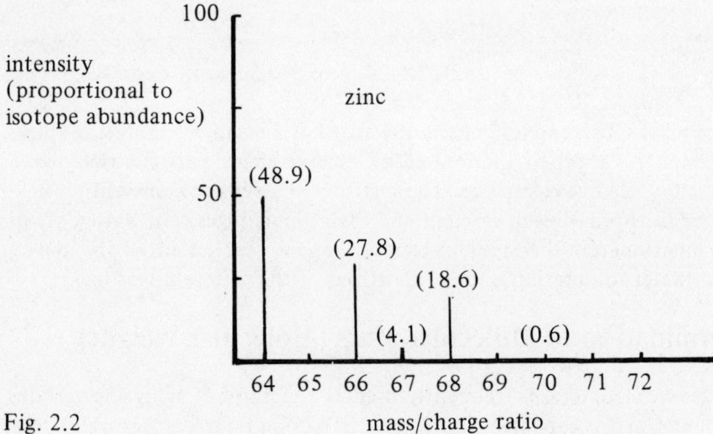

Fig. 2.2

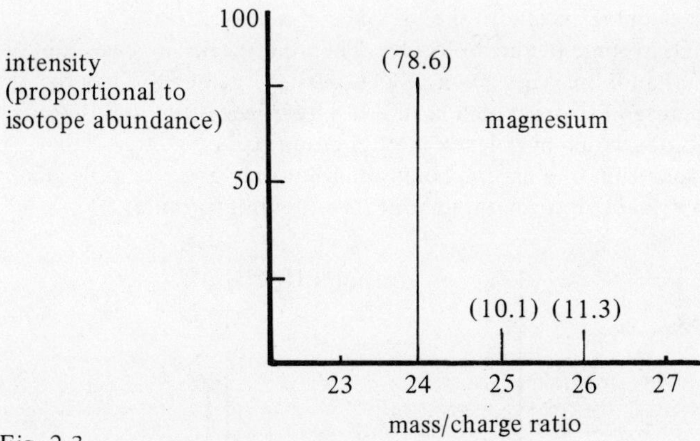

Fig. 2.3

Calculation of Atomic Mass (Atomic Weight)

From the values of percentage abundance, the atomic mass, or *average* mass of the atoms, of an element is readily calculated.

atomic mass = Σ (Fraction of atomic mass due to each isotope)
(Σ = 'The sum of')

e.g. for magnesium (Fig. 2.3) = $^{24}_{12}Mg$ 78.6%, $^{25}_{12}Mg$ 10.1%, $^{26}_{12}Mg$ 11.3%.

$$\text{atomic mass of Mg} = \frac{78.6 \times 24}{100} + \frac{10.1 \times 25}{100} + \frac{11.3 \times 26}{100}$$

$$= 18.864 + 2.525 + 2.938$$

$$= 24.327 = 24.3, \text{ to 3 significant figures}$$

The mass spectrometer is used nowadays as a means of analysing alloys. The sample is vaporised and ionised by striking an arc between two pieces of the alloy used as electrodes. The instrument gives the composition in terms of isotopes of each element and their percentages. The dating of rocks mentioned in the previous chapter depends on the use of the mass spectrometer to determine the proportions of the different isotopes.

Determination of Molecular Mass (Molecular Weight)

The mass spectrometer is frequently used as a method of analysing carbon compounds, and is capable of producing in half-an-hour a structural formula which would have previously taken days to produce. Fig. 2.4 shows part of the mass spectrum which could be obtained from a sample of ethanol. The peak at 46 corresponds to the molecule less one electron. A very small peak at 47 is due to the presence of a heavier isotope, i.e. ^{2}H, ^{13}C or ^{17}O, in some of the molecules. The other peaks are caused by molecules of ethanol breaking down to give smaller fragments. Thus the peak at 45 is caused by ions which have lost a hydrogen atom and the peak at 31 by ions which have lost a methyl group (CH_3).

The mass spectrum of a carbon compound not only gives its molecular mass but also valuable information about its structural formula.

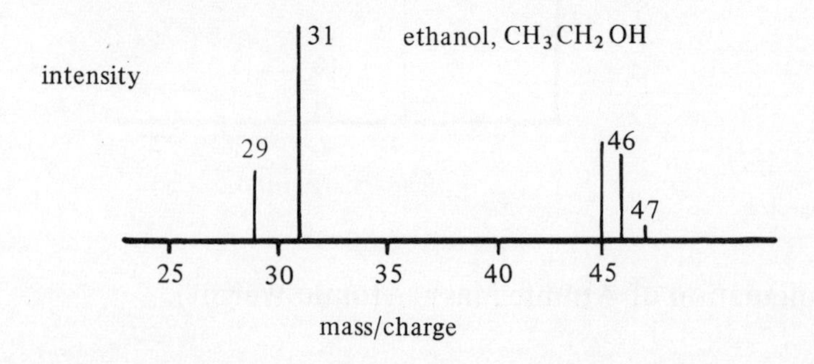

Fig. 2.4

Standards of Atomic Mass

Originally, the standard of atomic mass was the average mass of the hydrogen atom, namely, 1, since hydrogen was the lightest known element and presumably contained the lightest atoms. However, since hydrogen does not combine readily with many other elements, which would have simplified the determination of atomic masses by chemical means, atomic masses were in practice determined against other elements, which were themselves determinable against hydrogen. To simplify matters, oxygen, which combines readily with most elements and has an atomic mass of nearly 16 on the hydrogen scale, was taken as being exactly 16. Another result of this change of standard was that the atomic masses of many elements were nearer whole numbers.

However, it was later discovered that oxygen is a mixture of isotopes (as are most other elements, thus explaining fractional atomic masses) and so O = 16 referred to the average mass of the oxygen atoms in a normal sample. Although this did not affect chemists, physicists were beginning to use the mass spectrometer, and taking the lightest oxygen isotope as exactly 16. Hence, a 'chemical' and a 'physical' atomic mass scale arose.

In 1960 international agreement was achieved and it was decided to use the carbon-12 isotope as the standard of atomic masses, i.e. the mass of a ^{12}C atom is taken as exactly 12. Since most atomic mass determinations are now the result of mass number and isotope abundance determinations by mass spectrometer, it is more sensible to use a standard isotope rather than an average value for the mixture of isotopes in an element.

Some of the most modern mass spectrometers are accurate to greater than one part in 10 000. On the ^{12}C scale other isotopes can be shown to have masses which are not exactly integral values of atomic mass units (a.m.u.), e.g. the isotopic mass for ^{16}O is 15.9949. This enables the mass spectrometer to distinguish between molecules such as N_2, CO, and C_2H_4, each of which nominally has a molecular mass of 28.

Examples for practice.

Calculate the atomic masses of the following:

1. zinc, using the data given on Fig. 2.2,
2. silver, given its isotopic composition – 51.4% ^{107}Ag, 48.6% ^{109}Ag,
3. lead, given its isotopic composition – 1.5% ^{204}Pb, 23.6% ^{206}Pb, 22.6% ^{207}Pb, 52.3% ^{208}Pb.

Note: This last set of values may vary slightly from sample to sample, an unusual phenomenon for an element. This is because lead is the end product of several radioactive decay series and different series produce different isotopes.

3 The Mole

Introduction

Elements

The atomic mass of an element is the average mass in atomic mass units of the atoms of an element, according to its normal isotopic composition. However, when chemical reactions occur, large numbers of atoms are involved and it is impossible to measure such small quantities as a.m.u.

Let us consider the following examples:

	carbon		calcium		copper
mass of 1 atom (a.m.u.)	12		40		64
mass of x atoms (a.m.u.)	$12x$		$40x$		$64x$
ratio of masses of same number of atoms	12	:	40	:	64

Therefore, if we take 12 g of carbon, 40 g of calcium and 64 g of copper, each of these quantities will contain the same number of atoms. This number is very large and is called **Avogadro's number.** It is given the symbol N_A, and has the value 6.02×10^{23}. This number is also known as the **mole.**

The weight of one mole of atoms of an element is the atomic mass of the element expressed in grammes.

The number of moles in a given weight = $\dfrac{\text{given weight}}{\text{weight of 1 mole}}$

Covalent substances

When considering a covalent substance we can regard the molecular mass as being equivalent to the atomic mass of an element. For example:

1 mole of hydrogen, H_2, weighs 2 g. This contains two moles of atoms.
1 mole of carbon dioxide, CO_2, weighs 44 g.
1 mole of glucose, $C_6H_{12}O_6$, weighs 180 g.

Each of these quantities contain Avogadro's number of *molecules.* Thus, one mole of carbon dioxide contains N_A molecules, made up of N_A atoms of carbon and 2 N_A atoms of oxygen, i.e. 3 N_A atoms altogether.

The weight of one mole of molecules of a covalent substance is the molecular mass expressed in grammes.

Ionic compounds

For ionic compounds, formula mass is used instead of molecular mass, e.g. 1 mole of sodium chloride, NaCl, weighs 58.5 g. This quantity contains 1 mole of Na^+ ions and 1 mole of Cl^- ions, i.e. 2 N_A ions altogether.

Similarly, 1 mole of copper (II) nitrate, $Cu(NO_3)_2$, weighs 188 g and contains 1 mole of Cu^{2+} ions and 2 moles of NO_3^- ions, i.e. 3 N_A ions altogether.

Summary

We can define the mole in a number of ways as follows.
The mole is:

the number of atoms in the atomic mass in grammes of an element,
" " " molecules " molecular mass " " " " "
" " " " " " " " " a covalent compound,
" " " 'formulae' " formula mass " " an ionic compound.

By usage the mole, as well as being the *number* of particles, has come to mean also the total *mass* of those particles. Thus we can say that 27 g of aluminium *is* one mole of aluminium or that it *contains* one mole of aluminium.

Some examples of the meaning of 'mole' are given below.

1 mole of magnesium atoms are contained in 24 g of magnesium,
1 " " oxygen atoms " " " 16 g oxygen*,
1 " " oxygen molecules " " " 32 g oxygen*,
1 " " water molecules " " " 18 g water,
1 " *each* of K^+ and I^- ions " " " 166 g potassium iodide.

*It is essential to know precisely which type of particles is being considered.

Worked examples

1. How many moles of hydrogen atoms are contained in 6.4 g of methane? How many methane molecules are present?

Methane has the molecular formula CH_4 and its molecular mass is 16.

$\therefore$ 6.4 g methane contains $\frac{6.4}{16}$ moles of CH_4 molecules = 0.4 moles.

But, each methane molecule contains 4 hydrogen atoms.

$\therefore$ 6.4 g methane contains 0.4×4 moles of hydrogen atoms
= 1.6 moles of hydrogen atoms

No. of methane molecules in 6.4 g = 0.4 moles = $0.4 \times N_A$
$= 0.4 \times 6 \times 10^{23}$
$= 2.4 \times 10^{23}$

2. What is the weight of 0.1 mole of anhydrous sodium sulphate? How many ions does it contain?

Anhydrous sodium sulphate has the formula Na_2SO_4, and a formula mass of 142.

$\therefore$ 0.1 mole of anhydrous sodium sulphate weighs 0.1×142 g = 14.2 g

Sodium sulphate is in fact $(Na^+)_2SO_4^{2-}$, i.e. 1 mole of Na_2SO_4 contains 2 moles of Na^+ ions and 1 mole of SO_4^{2-} ions, a total of 3 moles of ions.

$\therefore$ 0.1 moles of Na_2SO_4 contains 0.3 moles of ions.

$\therefore$ No. of ions present = $0.3 \times N_A = 0.3 \times 6 \times 10^{23} = 1.8 \times 10^{23}$

Calculations involving Percentage Composition and Empirical Formulae

If we know the formula of a compound and the atomic masses of the elements present, it is a simple matter to work out the percentage of each element by mass.

Worked example

3. Nitric acid, HNO_3, has a formula mass of 1 + 14 + 48 = 63. What is its percentage composition by mass?

% Hydrogen = $\frac{1}{63} \times 100 = 1.58$ %   % Nitrogen = $\frac{14}{63} \times 100 = 22.22$ %

% Oxygen = $\frac{48}{63} \times 100$ or $100 - (1.58 + 22.22) = 76.20$ %

However, the reverse type of calculation is usually of greater importance. If a compound can be analysed and its percentage composition determined, it is then possible to obtain its empirical formula.

Worked example

4. The percentage composition by mass of nitric acid is – hydrogen 1.58%, nitrogen 22.22%, and oxygen 76.20%. What is its empirical formula?

The usual method is as follows:

Table 3.1

element	% by mass	$\frac{\%\ \text{mass}}{\text{atomic mass}}$	= moles of atoms of element	relative no. of moles
H	1.58	$\frac{1.58}{1}$	= 1.58	1
N	22.22	$\frac{22.22}{14}$	= 1.58	1
O	76.20	$\frac{76.20}{16}$	= 4.76	3

The figures in the last column are obtained by dividing each of the answers in the preceding column by the smallest answer there, in this case 1.58.

This type of calculation gives the *simplest* ratio of moles of atoms present in the compound and is known as the **empirical formula.** The empirical formula, therefore, of nitric acid is HNO_3. This is also its molecular formula. Sometimes a compound has a molecular formula that is different from its empirical formula, e.g. ethane has an empirical formula CH_3 and molecular formula C_2H_6. In such cases some extra information, apart from the percentage composition, is required to obtain the molecular formula. This is usually the molecular mass of the compound. In the case of ethane, the molecular mass is 30 and the empirical formula mass is 15, hence its molecular formula is twice its empirical formula.

Calculations from Equations

It is important from a practical viewpoint to be able to calculate the quantities of materials involved in chemical reactions. The key to this type of calculation is the chemical equation, which is an abbreviated statement of what is reacting and what is being produced in a chemical reaction. The equation must be balanced so that it shows the corresponding number of moles of reactants and products.

Worked example

5. What weight of iron can be obtained from 40 g of iron (III) oxide by reducing in a stream of hydrogen?

Balanced equation:

$$\underset{\text{1 mole}}{Fe_2O_3} + 3\,H_2 \longrightarrow \underset{\text{2 moles}}{2\,Fe} + 3\,H_2O$$

160 g iron (III) oxide can give 2×56 g iron = 112 g iron

∴ 1 g ’ ’ ’ ’ $\frac{112}{160}$ g ’

∴ 40 g ’ ’ ’ ’ $\frac{112 \times 40}{160}$ g iron = 28 g

Reactions in Solution

Many chemical reactions take place in solution. To be able to deal with them quantitatively we need to have a standard method of measuring concentrations. We do this in terms of molarities.

A **molar solution** (called M) contains 1 mole of *solute* per litre of *solution.*

For example:

1 litre of	M	sulphuric acid contains	1 mole (98 g)	of pure sulphuric acid	
½ ’ ’	M	’ ’ ’	½ mole (49 g)	’ ’ ’ ’	
1 ’ ’	0.5M	’ ’ ’	½ mole (49 g)	’ ’ ’ ’	
½ ’ ’	0.5M	’ ’ ’	¼ mole (24.5 g)	’ ’ ’ ’	
2 ’ ’	2M	’ ’ ’	4 moles (392 g)	’ ’ ’ ’	

The number of moles of solute in a given volume of solution
= volume (in litres) × molarity.

By titration it is possible to obtain the molarity of a solution. The method of calculation is given below.

Worked example

6. In an experiment 10 cm^3 of a solution of sulphuric acid neutralised 25 cm^3 of 0.1M NaOH solution. What was the molarity of the sulphuric acid?

Balanced equation: $2\ NaOH + H_2SO_4 \longrightarrow Na_2SO_4 + 2\ H_2O$

From the equation: 2 moles NaOH $\equiv$ 1 mole H_2SO_4

i.e. 2 litres M NaOH $\equiv$ 1 litre M H_2SO_4

In theory, 25 cm^3 0.1M NaOH require 12.5 cm^3 0.1M H_2SO_4

But in expt. 25 cm^3 0.1M NaOH require 10 cm^3 of H_2SO_4 of unknown molarity x.

These two quantities of acid can be equated i.e. 10 cm^3 of molarity x $\equiv$ 12.5 cm^3 0.1M

$x = \frac{12.5}{10} \times 0.1\text{M} = 0.125\text{M}$

Molarity of the sulphuric acid solution is 0.125M

The Mole and Gas Volumes

We have seen how we can relate a mole of material to an actual weight. However, in the case of gases it is often more useful to deal with volumes. Let us consider the following examples.

Table 3.2

gas	H_2	He	N_2	O_2	CO_2	SO_2
density in gl^{-1} at $0^\circ C$ and 1 atmos. pressure	0.0899	0.179	1.25	1.43	1.98	2.93
mass of 1 mole (g)	2.02	4	28	32	44	64.1
volume occupied by 1 mole of gas in litres (= mass/density)	22.46	22.36	22.40	22.38	22.22	21.88

If we weigh equal volumes of various gases, not surprisingly we find that the weights differ, since the densities of the gases are different. In the examples quoted above we can see that the density of a gas increases as the molecular mass increases. However, if we work out by proportion the volume that is occupied by one mole of each gas, then we observe that, within small limits, one mole of each gas always occupies the same volume provided that it is measured at the same temperature and pressure.

The volume occupied by one mole of *any* gas is 22.4 litres at Normal or Standard Temperature and Pressure (NTP or STP), that is at $0^\circ C$ or 273 K and 1 atmosphere pressure or 760 mm of mercury. At other temperatures and pressures this volume of course changes, in accordance with the gas laws, but we need not take account of that at this stage.

It may seem at first surprising that a mole of any gas occupies the same volume provided that temperature and pressure are constant. It is certainly not true for solids and liquids. In a gas, however, the molecules are relatively far apart so that the volume does not depend on the size of the actual molecules. Calculations show that in a gas, the molecules only occupy about 0.1 % of the total volume. The remainder is empty space.

Worked examples

7. What volume of carbon dioxide at STP would be produced by the action of excess dilute hydrochloric acid on 20 g of calcium carbonate?

Balanced equation:

$$CaCO_3 + 2\,HCl \longrightarrow CaCl_2 + CO_2 + H_2O$$

1 mole   1 mole

100 g $CaCO_3$ will produce 22.4 litres CO_2 at STP

$\therefore$ 1 g ’ ’ ’ $\frac{22.4}{100}$ ’ ’ ’ ’

$\therefore$ 20 g ’ ’ ’ $\frac{22.4 \times 20}{100}$ litres CO_2 at STP

$= 4.48$ litres

8. What volume of carbon dioxide at STP would be produced by the action of 200 cm^3 of M HCl on 20 g calcium carbonate?

At first sight, this calculation seems to be a repeat of example 7, but this time the amount of acid is also specified. In this problem it is essential to find out which reagent is in excess, since the volume of gas will depend on the amount of the other reagent.

Balanced equation: $CaCO_3 + 2\,HCl \longrightarrow CaCl_2 + CO_2 + H_2O$

1 mole   2 moles   1 mole

100 g $CaCO_3$ will react with 2 litres M HCl to give 22.4 litres CO_2 at STP

$\therefore$ 20 g ’ ’ ’ ’400 cm^3 M HCl ’ ’ $\frac{22.4}{5}$ ’ ’ ’ ’

We only have 200 cm^3 M HCl, so that the $CaCO_3$ is present in excess. Therefore, the amount of acid will decide the volume of gas obtained.

200 cm^3 M HCl will produce $\frac{22.4}{10}$ litres of CO_2 at STP = 2.24 litres.

Avogadro’s Law

Avogadro’s Law states that ‘equal volumes of all gases under the same conditions of temperature and pressure contain the same number of molecules’. This follows directly from what we considered in the previous section, since the volume of one mole of any gas is 22.4 litres at STP and, therefore contains Avogadro’s number of molecules.

Let us illustrate this with the following examples

Table 3.3

gas	vol. of 1 mole at STP	no. of molecules in 1 mole	no. of molecules in 1 litre
H_2	22.4 litres	N_A	$N_A/22.4$
CH_4	22.4 ”	N_A } = 6×10^{23}	$N_A/22.4$ } = 2.7×10^{22}
CO_2	22.4 ”	N_A	$N_A/22.4$

Worked examples

9. Silver (I) nitrate decomposes on heating to form silver metal, nitrogen dioxide and oxygen. What is the percentage composition by volume of the gas mixture produced?

Balanced equation: $2\,AgNO_3 \longrightarrow 2\,Ag + 2\,NO_2 + O_2$
2 moles → 2 moles NO_2, 1 mole O_2

Ratio of gases by moles, $NO_2 : O_2$ = 2 moles : 1 mole
” ” ” volume, $NO_2 : O_2$ = 2 vols. : 1 vol

This follows from Avogadro's Law, since the gas mixture produced in the above reaction contains NO_2 and O_2 at the same temperature and pressure.

% NO_2 = 66.7 % and % O_2 = 33.3 %

10. What volume of oxygen is required to burn completely 20 cm^3 of ethene? What volume of carbon dioxide would be produced? All volumes are measured at the same temperature and pressure.

	C_2H_4	+ $3\,O_2$ ⟶	$2\,CO_2$	+ $2\,H_2O$
Balanced equation:	1 mole	3 moles	2 moles	
By Avogadro's Law	1 vol.	3 vols.	2 vols.	
	20 cm^3	3×20 cm^3	2×20 cm^3	

Volume of oxygen required = 60 cm^3
Volume of CO_2 produced = 40 cm^3

11. One volume of propane is mixed with 10 volumes of oxygen and the mixture is exploded. What is the volume composition of the resulting gas mixture? All volumes are measured at the same conditions of room temperature and pressure.

Balanced equation:	C_3H_8	+ $5\,O_2$	$\longrightarrow$ $3\,CO_2$	+ $4\,H_2O$
	1 mole	5 moles	3 moles	4 moles
By Avogadro's Law	1 vol.	5 vols.	3 vols.	(4 vols. if the water is above its boiling point)

In this calculation, since we are working at room temperature, we can neglect the water.

1 volume of propane reacts with 5 volumes of O_2 to produce 3 volumes of CO_2.
5 volumes of oxygen remain unused.
Volume composition of mixture after explosion is $CO_2 : O_2 = 3 : 5$
or $CO_2 = 37.5\ \%$ and $O_2 = 62.5\ \%$

Quantitative Electrolysis

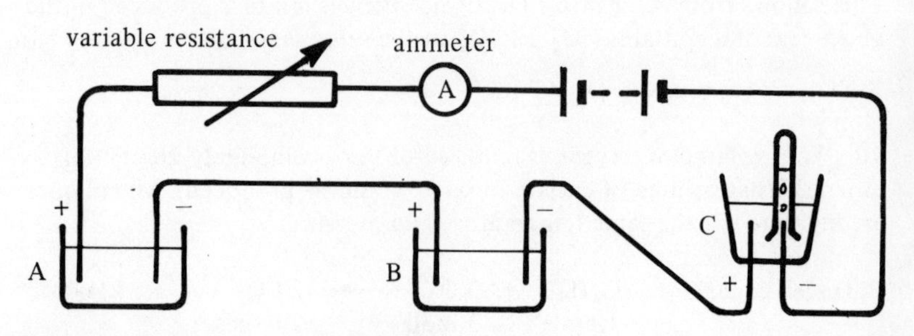

Fig. 3.1

Three cells containing respectively silver (I) nitrate solution, copper (II) sulphate solution and dilute sulphuric acid are connected in series as shown in Fig. 3.1. Let us suppose that in a particular experiment a current of one amp is passed through the circuit for 965 seconds. Relevant data and results are shown in table 3.4.

Table 3.4

cell	A	B	C
solution	silver(I) nitrate	copper(II) sulphate	sulphuric acid
type of electrodes	silver	copper	platinum or carbon
current used (A)	1.0	1.0	1.0
time for which current passed (s)	965	965	965
quantity of electricity used = current × time (coulombs)	965	965	965
cathode product	silver	copper	hydrogen
cathode reaction	$Ag^+ + e \rightarrow Ag$	$Cu^{2+} + 2e \rightarrow Cu$	$2H^+ + 2e \rightarrow H_2(g)$
amount of cathode product obtained	1.08 g	0.32 g	0.01 g (0.112 litres at STP)
mass of 1 mole of cathode product	108 g	64 g	2 g (22.4 litres at STP)
no. of moles of cathode product obtained	0.01	0.005	0.005
quantity of electricity required to produce 1 mole of cathode product (coulombs)	$\frac{965}{0.01}$ = 96 500	$\frac{965}{0.005}$ = 193 000	$\frac{965}{0.005}$ = 193 000

In general, it is discovered that the production of 1 mole of an element by electrolysis always requires nF coulombs, where n is the charge of the ion being discharged, and F is a quantity of electricity known as the **Faraday** whose value is 96 500 coulombs. In cell C in Fig. 3.1, we are considering the production of 1 mole of hydrogen gas, and as this requires the discharge of two moles of hydrogen ions at the cathode, 2 × 96 500 coulombs or 2 Faradays are required.

Thus, to discharge

1 mole H atoms	(1 g)	from 1 mole H^+	needs	96 500 coulombs	or 1 F
1 ' H_2 molecules	(2 g)	' 2 moles H^+	'	2 × 96 500 '	' 2 F
1 ' Ag atoms	(108 g)	' 1 mole Ag^+	'	96 500 '	' 1 F
1 ' Cu '	(64 g)	' 1 ' Cu^{2+}	'	2 × 96 500 '	' 2 F
1 ' Al '	(27 g)	' 1 ' Al^{3+}	'	3 × 96 500 '	' 3 F

We can make use of these results to find a value for Avogadro's number, N_A. 1 mole of copper is deposited by 2 × 96 500 coulombs.

Cathode reaction: $Cu^{2+} + 2e \longrightarrow Cu$

i.e. 1 mole of copper ions + 2 moles of electrons give 1 mole of copper atoms

i.e. N_A Cu^{2+} + 2 N_A e give N_A Cu atoms

i.e. 2 N_A electrons are equivalent to 2 × 96 500 coulombs (2 Faradays).

∴ N_A electrons ≡ 96 500 coulombs ≡ 1 Faraday

A Faraday, therefore, is the total charge due to a mole, or Avogadro's number, of electrons. To obtain a value for Avogadro's number we need to divide this total charge by the charge on one electron. This can be determined by what is known as Millikan's oil drop experiment. The charge on an electron has been found to be 1.6×10^{-19} coulombs.

$$\text{Avogadro's No., } N_A = \frac{\text{1 Faraday}}{\text{charge on 1 electron}} = \frac{96\,500}{1.6 \times 10^{-19}}$$

$$= 6.03 \times 10^{23}$$

Worked example

12. A solution of copper(II) chloride is electrolysed between carbon electrodes using a current of 0.5 A for 64 minutes and 20 seconds. Give the ion-electron equations for the reactions at each electrode. Calculate (a) the weight of copper obtained and (b) the volume of chlorine, at STP, produced.

Current = 0.5 A   Time = 3860 s

Quantity of electricity used = 0.5 × 3860 coulombs = 1930 coulombs

(a) At the cathode: $Cu^{2+} + 2\,e \longrightarrow Cu$

i.e. 2 moles of electrons
2 Faradays
2 × 96 500 coulombs
} required to produce 1 mole of copper = 64 g

1930 coulombs produce $\frac{64 \times 1930}{2 \times 96\,500}$ g of copper = 0.64 g

(b) At the anode $2\,Cl^- \longrightarrow Cl_2 + 2\,e$

i.e. 2 moles of electrons
2 Faradays
2 × 96 500 coulombs } involved in producing 1 mole of chlorine molecules = 22.4 litres at STP.

1930 coulombs produce $\frac{22.4 \times 1930}{2 \times 96500}$ litres of chlorine at STP.

$= 0.224$ litres

Summary

The mole consists of a very large number of particles. It is also known as Avogadro's number, $N_A = 6.02 \times 10^{23}$. For most calculations 6×10^{23} is sufficiently accurate. The mole can be defined in each of the following different ways:

1. the number of atoms in the atomic mass in grammes of an element;
2. the number of molecules in the molecular mass in grammes of an element or compound;
3. the number of molecules in 22.4 litres of a gas at STP;
4. the number of electrons in a Faraday of electricity, i.e. in 96 500 coulombs;
5. the number of electrons needed to deposit the atomic mass in grammes of a monovalent element.

It is also frequently used to denote the *mass* of Avogadro's number of particles.

Examples for practice

Note: Do *not* use 'M' as an abbreviation for 'mole'. 'M' means 'molar' and refers only to solutions. The recognised abbreviation for 'mole' is 'mol'.

Atomic masses are usually taken to the nearest whole number, but chlorine is always taken as 35.5 and Copper is usually taken as 64.

1. How many atoms are there in the following?
(a) 16 g of sulphur
(b) 15.4 g of tetrachloromethane
(c) 2 g of water

2. How many molecules are there in the following?
(a) 9.5 g of fluorine
(b) 1.1 g of carbon dioxide
(c) 10 g of hydrogen
(d) 4.48 l of oxygen at STP
(e) 11.2 l of carbon dioxide at STP

3. How many ions are there in the following?
(a) 9.4 g of potassium oxide
(b) 14.8 g of calcium hydroxide
(c) 100 g of iron (III) sulphate

4. How many atoms are contained in a cube of side 10 cm of the element yttrium (At. Mass 89)? Each cm^3 of yttrium weighs 4.45 g.

5. What is the weight of a mole of each of the following?
(a) Germanium
(b) Iodine molecules
(c) Manganese (IV) oxide
(d) Copper sulphate crystals $CuSO_4 \cdot 5H_2O$

6. What is the weight of the following quantities?
(a) 1.5 moles of vanadium
(b) 2.5 moles of nitrogen molecules
(c) 0.1 moles of nitric acid
(d) 0.5 moles of iron (III) sulphate

7. How many moles are there in the following?
(a) 4 g of sulphur
(b) 980 g of sulphuric acid
(c) 40 g of bromine molecules
(d) 40 g of neon

8. How many moles of **carbon** are there in the following?
(a) 7 g of carbon monoxide
(b) 92 g of ethanol C_2H_5OH
(c) 13 g of benzene C_6H_6
(d) 21 g of propene C_3H_6

9. Find the percentage composition of the elements in
(a) Ethanol C_2H_5OH
(b) Iron(III) nitrate
(c) Copper(II) sulphate (anhydrous)
(d) Phosphoric acid

10. Which of the fertilisers urea $CO(NH_2)_2$ or ammonium nitrate would, weight for weight, provide the better supply of nitrogen?

11. Find the empirical formula of each of the following compounds given their percentage **compositions**:

(a) H 2.1%, S 32.6%, O 65.3%
(b) C 76.5%, H 6.4%, O 17.1%
(c) Mg 9.8%, S 13%, O 26%, H_2O 51.2%
(d) Pb 62.5%, N 8.5%, O 29%

12. A compound consists of 85.7% carbon and 14.3% hydrogen. Its molecular mass is 28, what is its molecular formula?

13. Find the empirical formula of a compound, 3.27 g of which contains 1.04 g of potassium, 0.95 g of chlorine and 1.28 g of oxygen. (There is no need to work out the % composition first).

14. How many grams of calcium chloride would be required to precipitate 14.35 g of silver(I) chloride from excess silver(I) nitrate solution? (As with all calculations of this type begin by writing a balanced chemical equation).

15. What weight of lead oxide will be produced by heating 6.62 g lead nitrate which decomposes according to the following equation?

$$2Pb(NO_3)_2 \longrightarrow 2PbO + 4\ NO_2 + O_2$$

16. What weight of zinc will completely dissolve in 100 cm^3 of 2M sulphuric acid?

17. What weight of calcium carbonate will dissolve completely in 50 cm^3 M hydrochloric acid?

18. How many moles of solute are there in each of the following solutions?
(a) 500 cm^3 5M
(b) 100 cm^3 0.2M
(c) 2 l 0.4M
(d) 250 cm^3 0.8M

19. Calculate the weight of solute in each of the following:
(a) 1 l 0.2M sulphuric acid
(b) 5 l 4M potassium bromide
(c) 600 cm^3 0.1M silver(I) nitrate
(d) 200 cm^3 M sodium carbonate

20. Calculate the molarity of each of the following solutions:
(a) 49 g of sulphuric acid in 2 l of solution
(b) 21 g of nitric acid in 250 cm^3 of solution
(c) 3.31 g of lead (II) nitrate in 200 cm^3 of solution
(d) 530 g of sodium carbonate in 10 l of solution

21. How much 0.5M HCl is required to neutralise the following?
(a) 10 cm^3 M Na_2CO_3 solution
(b) 25 cm^3 0.2M KOH solution
(c) 10 cm^3 M KOH solution

22. 25 cm^3 of ammonium hydroxide neutralised completely 10 cm^3 of 0.5M sulphuric acid. What was the molarity of the ammonium hydroxide?

23. Calculate the densities (in g/litre) of the following gases at STP:
(a) CH_2F_2
(b) ammonia
(c) neon

24. Here are data concerning a small gas cylinder.
Capacity: 100 ml, Temperature: 0 °C, Pressure of gas *X* inside cylinder: 4 atmospheres, Wt. of cylinder when evacuated: 80.0 g, Wt. of cylinder with gas *X*: 80.5 g.
Calculate the approximate molecular weight of the gas *X*. (S.C.E.E.B.)

25. Calculate the volume of gas at STP produced in the following reactions. Assume all reactions to go to completion.
(a) 21 g of magnesium carbonate decomposed by heat.
(b) 13 g of zinc added to excess dilute hydrochloric acid.
(c) 13 g of zinc added to 100 cm^3 2M hydrochloric acid.

26. What weight of water would be produced by burning 11.2 litres of hydrogen (measured at STP) in excess oxygen?

27. What weight of copper sulphide would be precipitated if 448 cm^3 at STP of H_2S were passed into excess copper nitrate solution?

$$Cu(NO_3)_2 + H_2S \longrightarrow CuS + 2HNO_3$$

28. (a) What is the weight of lead sulphide which could be obtained by passing hydrogen sulphide gas, H_2S through a solution of 32.5 g of lead acetate, $(CH_3COO)_2Pb$, in 1 litre of water? (b) What volume of hydrogen sulphide at STP would be needed for the reaction? (c) What would be the molarity of the acetic acid remaining, assuming that no volume change had occurred during the reaction? (S.C.E.E.B.)

29. For each of the following gases, (i) write down the equation for its complete combustion, and (ii) calculate the volume of oxygen required for the complete combustion of 1 litre of it and also the volume of carbon dioxide produced.

(a) methane (b) ethene (c) methanal (formaldehyde) – CH_2O
All volumes are measured under the same conditions of temperature and pressure.

30. Given that 1 Faraday = 96 500 coulombs, what weight of silver will be deposited on passing a current of 2 amps through silver(I) nitrate solution for 30 minutes?

31. During an electrolysis 0.25 g of hydrogen are collected. What weight of copper would be collected if the same current is used for the same time in electrolysis of copper (II) chloride solution?

32. What volumes of gases at STP would be produced by the electrolysis of sodium chloride solution using carbon electrodes, if a current of 1 amp is passed for 15 minutes?

33. A current of 10 A is passed through molten magnesium chloride for 15 minutes. Give the equations for the reactions that take place at both electrodes. Calculate the weight of product formed at the cathode. (S.C.E.E.B.)

4 Energy and Chemical Reactions

Throughout your study of chemistry, you will have frequently observed that when chemical reactions occur they are accompanied by a significant change in energy. Most of the reactions encountered will have involved a release of energy, usually in the form of heat and are thus said to be **exothermic.** Examples of such reactions include

oxidation of metals and non-metals,
combustion of carbon compounds,
neutralisation of acids by alkalis and metals,
displacement of less reactive metals,
precipitation of insoluble salts.

Reactions in which heat is absorbed are said to be **endothermic.** Though less frequent, such reactions do take place, e.g. the dissolving of certain salts (e.g. NH_4NO_3, KNO_3) in water and the production of water gas from the reaction of steam on hot coke.

$$C_{(s)} + H_2O_{(g)} \longrightarrow \underbrace{CO_{(g)} + H_{2\,(g)}}_{\text{water gas}}$$

Energy may also be released in a chemical reaction in other forms, namely, as light (e.g. magnesium 'flare' produced when it burns) or sound (e.g. hydrogen – oxygen and hydrogen – chlorine explosions).

Enthalpy Change

When an exothermic reaction occurs, the reactants release heat to the surroundings and the products possess less energy as a result. The difference in energy between reactants and products is called the heat of reaction or **enthalpy change** (symbol: ΔH). Since, in this case, energy is lost by the reactants, ΔH is said to be negative (Fig. 4.1). Nowadays, ΔH values are usually quoted in kilojoules per mole of reactant or product, abbreviated to kJ mol^{-1}. In an endothermic reaction, the reactants take in heat from the surroundings, so that the products possess more energy than the reactants. Since there is an energy gain by the products, ΔH is positive

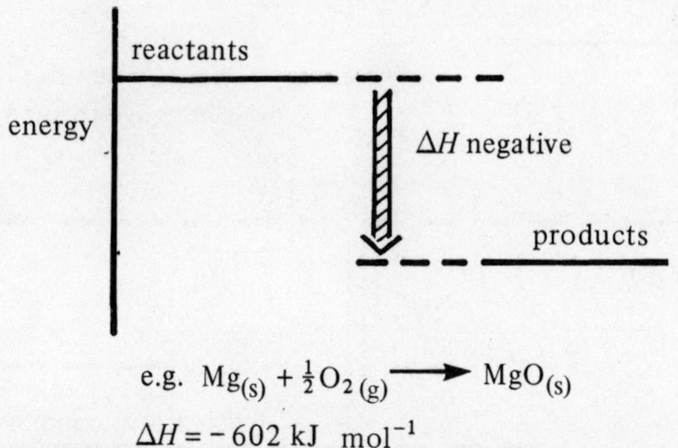

Fig. 4.1 An exothermic reaction

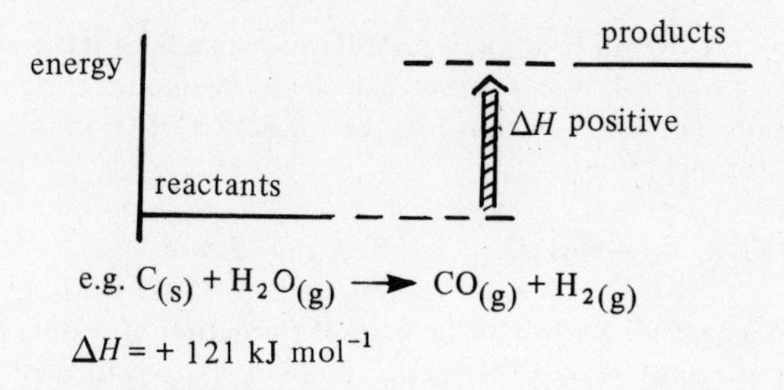

Fig. 4.2 An endothermic reaction

The enthalpy change for a reaction may be regarded as the net result of several intermediate stages, each of which involves a change in energy. To illustrate this point consider the reaction between hydrogen and oxygen (Figs. 4.3 and 4.4).

$$2\,H_{2\,(g)} + O_{2\,(g)} \longrightarrow 2\,H_2O_{(l)} \qquad \Delta H = -572 \text{ kJ}$$

Fig. 4.3 energy is needed to break these bonds (endothermic). Energy is released in forming new bonds (exothermic).

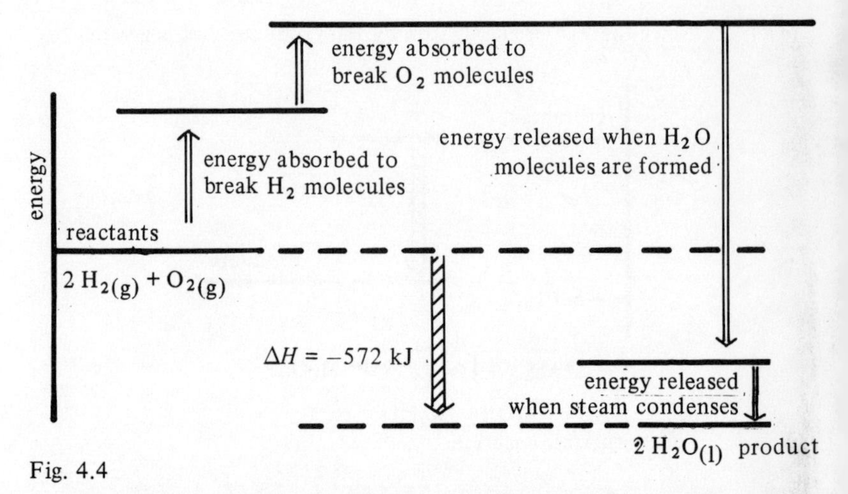

Fig. 4.4

In Fig. 4.4, the overall reaction is exothermic since the total amount of energy required to break bonds is less than the total amount released in bond-formation. The reaction concerns the formation of two moles of water. Thus, for one mole:

$$H_{2\,(g)} + \tfrac{1}{2}\,O_{2\,(g)} \longrightarrow H_2O_{(l)} \quad \Delta H = -\,286 \text{ kJ mol}^{-1}$$

This enthalpy change is known as the **heat of formation** of water, since it is the quantity of energy released when one mole of water is formed from its elements in their natural state. In general, the heat of formation of a compound is the change in enthalpy when one mole of the compound is formed from its elements in their natural state.

We can also consider the reaction given above from the point of view of one of the reactants. The equation tells us that the complete combustion of one mole of hydrogen requires half a mole of oxygen and the ΔH value informs us that 286 kJ of energy are released in the process. Consequently, the enthalpy change can also be regarded as the **heat of combustion** of hydrogen. Strictly speaking, this and other ΔH values are measured under specific conditions of temperature and pressure. These conditions are universally recognised and are 25°C or 298 K and one atmosphere pressure.

Measuring enthalpy changes

Note: For practical details, see appendix to this chapter, page 38.

1. Heat of neutralisation. When an acid and an alkali react, an exothermic reaction occurs in which hydrogen and hydroxyl ions combine to form water.

$$\underset{\text{from the acid}}{H^+_{(aq)}} + \underset{\text{from the alkali}}{OH^-_{(aq)}} \longrightarrow H_2O_{(l)}$$

When both acid and alkali are strong, i.e. they are both fully dissociated into ions, the heat of neutralisation is fairly constant at -57 kJ mol^{-1}. The other ions present are merely spectator ions. If either the acid or alkali is weak, i.e. only partially dissociated into ions, then the enthalpy change is less since energy is needed to complete the dissociation.

e.g. $CH_3COOH \longrightarrow CH_3COO^-_{(aq)} + H^+_{(aq)}$

Table 4.1 gives ΔH values which illustrate these points.

Table 4.1

acid	alkali	ΔH (kJ mol^{-1})
HCl	NaOH	-57
HNO_3	NaOH	-57
HCl	KOH	-57
CH_3COOH	NaOH	-55
HCl	NH_3 solution	-52

2. Heat of combustion. When we compare the heat of combustion of the members of a homologous series, e.g. the alcohols (table 4.2), we see that there is a fairly constant difference in ΔH between succeeding members. This is to be expected since there is a constant difference in chemical structure, namely the $-CH_2-$ group of atoms.

Table 4.2

alcohol	formula	ΔH (kJ mol^{-1})	difference in ΔH between succeeding members (kJ)
methanol	CH_3OH	-715	
			655
ethanol	CH_3CH_2OH	-1370	
			640
propan-l-ol	$CH_3CH_2CH_2OH$	-2010	
			660
butan-l-ol	$CH_3CH_2CH_2CH_2OH$	-2670	

i.e. $CH_3CH_2OH_{(1)} + 3O_{2(g)} \rightarrow 2CO_{2(g)} + 3H_2O_{(1)}$ $\quad \Delta H = -1370$ kJ mol^{-1}

3. Heat of solution. Enthalpy changes occur when ionic compounds are dissolved in water. These ΔH values are not usually high but are frequently endothermic. Table 4.3 gives a few examples of heats of solution.

Table 4.3

solute	ΔH (kJ mol^{-1})
NH_4NO_3	+ 26
KNO_3	+ 35
$Ca(NO_3)_2$	− 19

Hess's Law

The law of conservation of energy states that energy cannot be created or destroyed, but can only be converted from one form into another. When applying this law to chemical changes, the following statement can be made. The total enthalpy change in a chemical reaction depends only on the chemical nature and physical state of the initial reactants and final products, and is independent of any intermediate steps. This statement is known as Hess's law.

The following example is given to illustrate this law. Two routes are available for converting solid sodium hydroxide into a solution of sodium chloride.

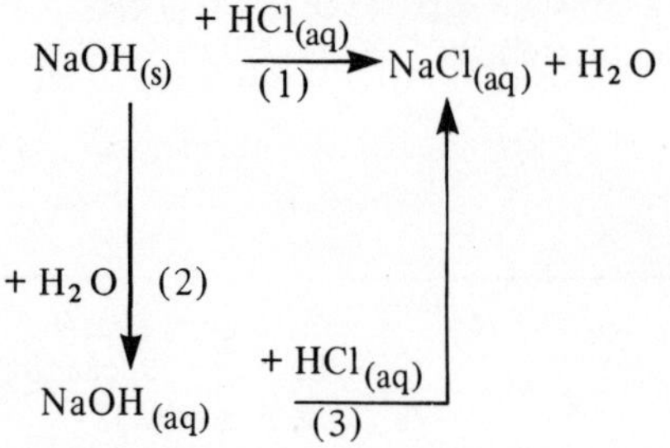

$$NaOH_{(s)} \xrightarrow[(1)]{+ HCl_{(aq)}} NaCl_{(aq)} + H_2O$$

$$NaOH_{(s)} \xrightarrow[(2)]{+ H_2O} NaOH_{(aq)} \xrightarrow[(3)]{+ HCl_{(aq)}} NaCl_{(aq)} + H_2O$$

1st route: Solid NaOH dissolved directly in dilute hydrochloric acid.
$\Delta H_{(1)} = -\ 100$ kJ mol^{-1}

2nd route. Solid NaOH dissolved first in water, then this solution neutralised with dilute acid.

$\Delta H_{(2)} = -\ 43$ kJ mol^{-1};
$\Delta H_{(3)} = -\ 57$ kJ mol^{-1}

$$\therefore \Delta H_{(1)} = \Delta H_{(2)} + \Delta H_{(3)}$$

Hess's law is important chiefly because it enables us to calculate enthalpy changes which are very difficult or impossible to measure by practical means and also to obtain energy values for the bonds which hold atoms together. In the following calculations we shall make use of the appropriate data given on page 202.

Application of Hess's Law

Finding the heat of formation of a compound.

$$C_{(s)} + 2\,H_{2\,(g)} \longrightarrow CH_{4\,(g)}$$

The above equation represents the formation of methane from its elements in their natural state. The ΔH value for this reaction, i.e. the heat of formation of methane, cannot be measured experimentally, but we can obtain its value using Hess's law. This involves devising an 'alternative route' with reactions whose ΔH values are known. This is shown below.

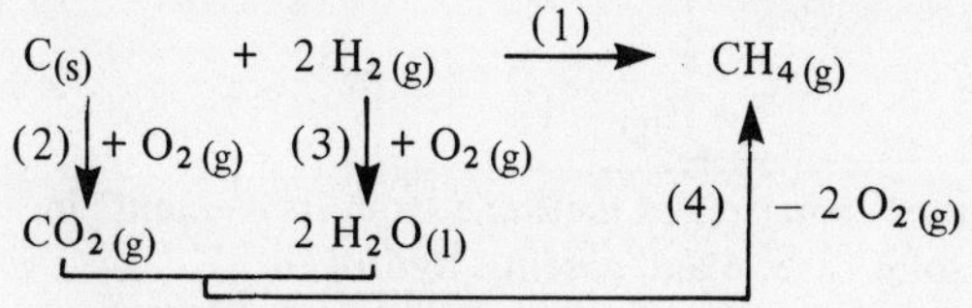

$$C_{(s)} \quad + \quad 2\,H_{2\,(g)} \xrightarrow{(1)} CH_{4\,(g)}$$

(2) $\downarrow + O_{2\,(g)}$ (3) $\downarrow + O_{2\,(g)}$ (4) $\uparrow - 2\,O_{2\,(g)}$

$CO_{2\,(g)}$ $2\,H_2O_{(l)}$

Reactions (2), (3) and (4) combine to provide the alternative route to reaction (1)

The known enthalpy changes are as follows:

(a) $C_{(s)} + O_{2\,(g)} \longrightarrow CO_{2\,(g)}$ Heat of combustion of carbon,

$\Delta H_{(2)} = -\,394\ \text{kJ mol}^{-1}$

(b) $H_{2\,(g)} + \frac{1}{2}O_{2\,(g)} \longrightarrow H_2O_{(l)}$ Heat of combustion of hydrogen,

$\Delta H = -\,286\ \text{kJ mol}^{-1}$

Reaction (3) involves 2 moles of hydrogen, $\Delta H_{(3)} = -572$ kJ.

(c) $CH_{4\,(g)} + 2\,O_{2\,(g)} \longrightarrow CO_{2\,(g)} + 2H_2O_{(l)}$ Heat of combustion of methane, $\Delta H = -\,890\ \text{kJ mol}^{-1}$

Reaction (4) is essentially the reverse of this.

i.e. $CO_{2\,(g)} + 2H_2O_{(l)} \longrightarrow CH_{4\,(g)} + 2O_{2\,(g)}$ $\Delta H_{(4)} = +\,890\ \text{kJ mol}^{-1}$

By Hess's Law $\Delta H_{(1)} = \Delta H_{(2)} + \Delta H_{(3)} + \Delta H_{(4)}$

$= -\,394 - 572 + 890 = -\,76$ kJ

i.e. Heat of formation of methane $= -\,76\ \text{kJ mol}^{-1}$

Finding the energy of covalent bonds. We can now make use of the heat of formation of methane to obtain an energy value for the carbon – hydrogen bond in methane. The formation of methane from its elements can be seen as a series of bond-breaking and bond-forming processes. The heat of formation is the algebraic sum of all of the bond breaking and bond-forming energies.

i.e. $\Delta H_f = \Sigma$ (bond-breaking energies) + Σ (bond-forming energies)

endothermic, hence + ve   exothermic, hence – ve

Methane:

$$C_{(s)} + 2H_{2(g)} \longrightarrow CH_{4(g)} \qquad \Delta H = -76 \text{ kJ mol}^{-1}$$

$C_{(s)}$	$2H_{2(g)}$	$CH_{4(g)}$
Bonds joining C atoms have to be broken	Bonds joining H atoms have to be broken	New C – H bonds have to be formed

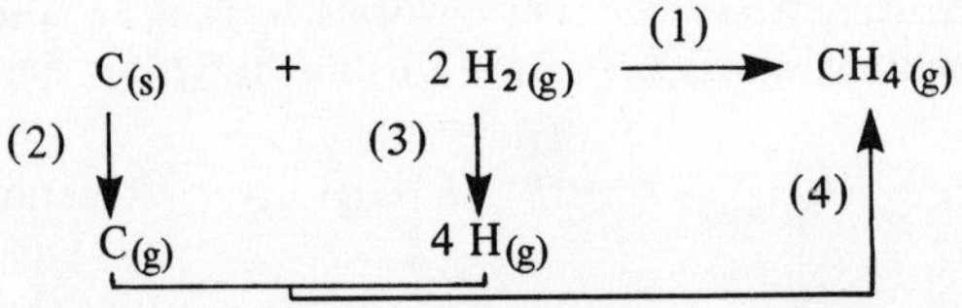

$$C_{(s)} + 2H_{2(g)} \xrightarrow{(1)} CH_{4(g)}$$

(2) $C_{(s)} \downarrow C_{(g)}$ (3) $2H_{2(g)} \downarrow 4H_{(g)}$ (4) $C_{(g)} + 4H_{(g)} \uparrow CH_{4(g)}$

Step (1) represents the formation of methane from its elements in their natural state, i.e. solid carbon and gaseous hydrogen.

$$\Delta H_{(1)} = -76 \text{ kJ mol}^{-1}.$$

Step (2) involves the breaking of all the bonds linking carbon atoms together in a mole of graphite so that the atoms become completely separate, i.e. in the gas phase. The energy required to do this is called the heat of sublimation of carbon.

$$C_{(s)} \longrightarrow C_{(g)} \qquad \Delta H_{(2)} = +715 \text{ kJ mol}^{-1}$$

Step (3) represents the dissociation of two moles of hydrogen molecules into completely separate atoms. The energy required for this is twice the heat of dissociation of hydrogen.

$$2H_{2(g)} \longrightarrow 4H_{(g)} \qquad \Delta H_{(3)} = +2 \times 435 \text{ kJ} = +870 \text{ kJ}$$

By Hess's law, $\Delta H_{(1)} = \Delta H_{(2)} + \Delta H_{(3)} + \Delta H_{(4)}$

$$-76 = +715 + 870 + \Delta H_{(4)}$$

$$\therefore \Delta H_{(4)} = -1661 \text{ kJ}$$

This is the energy released when one mole of methane molecules is formed from separate carbon and hydrogen atoms. Each molecule requires the formation of four C – H bonds. Hence the average energy released when a C – H bond is formed in methane is $-\dfrac{1661}{4} = -415 \text{ kJ mol}^{-1}$.

This value is very close to that quoted for the C – H bond in the table of mean bond dissociation energies. This table, given on page 202, shows the amount of energy (in kJ mol^{-1}) needed to break or dissociate various bonds. Each is a 'mean' value since it is the average over a wide variety of compounds in which that bond occurs.

Worked examples

Ethane, C_2H_6:

(a) Heat of formation, ΔH_f

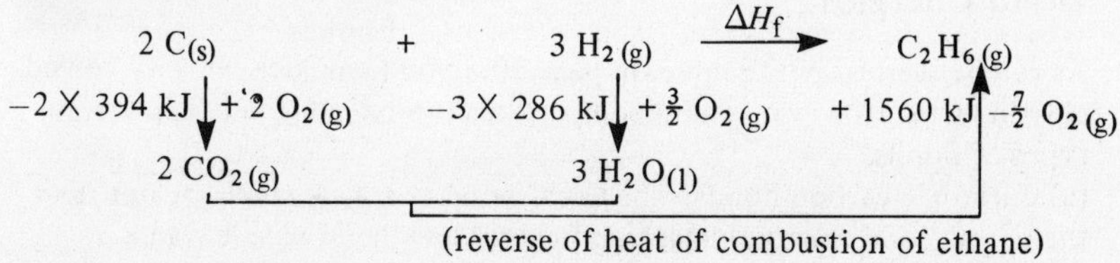

$$
\begin{array}{ccccc}
2\,C_{(s)} & + & 3\,H_{2\,(g)} & \xrightarrow{\Delta H_f} & C_2H_{6\,(g)} \\
-2 \times 394\text{ kJ} \downarrow +2\,O_{2\,(g)} & & -3 \times 286\text{ kJ} \downarrow +\tfrac{3}{2}\,O_{2\,(g)} & & +1560\text{ kJ} \uparrow -\tfrac{7}{2}\,O_{2\,(g)} \\
2\,CO_{2\,(g)} & & 3\,H_2O_{(l)} & &
\end{array}
$$

(reverse of heat of combustion of ethane)

By Hess's law, $\Delta H_f = -\ 788 - 858 + 1560$ kJ $= -\ 86$ kJ mol^{-1}

(b) C – C bond energy

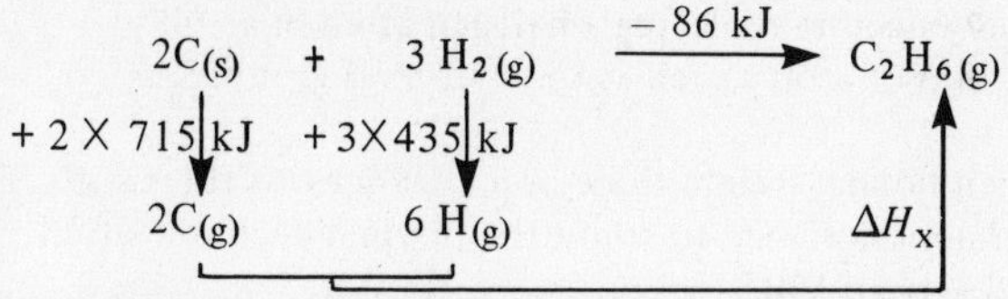

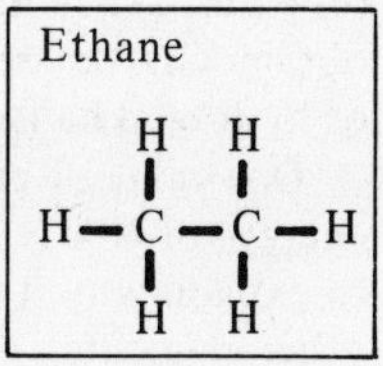

$$
\begin{array}{cccc}
2C_{(s)} & + \quad 3\,H_{2\,(g)} & \xrightarrow{-86\text{ kJ}} & C_2H_{6\,(g)} \\
+2 \times 715 \downarrow \text{kJ} & +3 \times 435 \downarrow \text{kJ} & & \uparrow \Delta H_x \\
2C_{(g)} & 6\,H_{(g)} & &
\end{array}
$$

Ethane

$$
\begin{array}{ccccc}
 & H & & H & \\
 & | & & | & \\
H - & C & - & C & - H \\
 & | & & | & \\
 & H & & H &
\end{array}
$$

By Hess's law, $-\ 86 = +\ 1430 + 1305 + \Delta H_x$

$\Delta H_x = -\ 2821$ kJ

This is the amount of energy released when one mole of ethane is formed from separate carbon and hydrogen atoms, i.e. when six moles of C – H bonds and one mole of C – C bonds are formed

– 2821 = energy released in forming 6 moles of C – H bonds + energy released in forming 1 mole of C – C bonds (X)

= – 6 × 414 + X

∴ X = – 337 kJ

∴ C – C bond energy = 337 kJ mol^{-1}

Carry out similar calculations for –

(i) Ethene, to find its heat of formation and the carbon-carbon double bond energy, and

(ii) Ethyne, to find its heat of formation and the carbon-carbon triple bond energy.
Use appropriate data from the tables given on page 202.

Structural formulae: Ethene –

```
H     H
 \   /
  C=C
 /   \
H     H
```

Ethyne – H—C≡C—H

Bond Energies

As can be seen from the table on page 202, bond energies can vary considerably in value. However, useful comparisons can be made between certain types of bonds.
(a) Carbon – carbon bonds: The C = C bond energy is much greater than the C – C bond energy, but is significantly less than twice its value. Similarly, the C ≡ C bond energy is less than three times that of the C – C. Try to explain these facts with reference to the electrons involved in bonding. This is dealt with in chapter 13.
(b) Carbon – halogen bonds: These bond energies have a bearing on the relative difficulty experienced in replacing a halogen atom in an alkyl halide with another group of atoms such as the hydroxyl group. See page 156.
(c) Hydrogen – halogen bonds: Relate these bond energies to the results obtained in the experiment designed to study the thermal decomposition of hydrogen halides. See page 100.

One value of particular interest not quoted on the table is the bond energy for N ≡ N. The very high value for this bond, 941 kJ mol^{-1}, helps to explain why it is so difficult to persuade nitrogen to combine with other elements.

Examples for Practice

Note: The heat of sublimation of carbon is 715 kJ mol^{-1}.

1. 1.00 g of ethanol was burned and the heat produced warmed 5 litres of water from 20.1 °C to 21.5 °C. Calculate the heat of combustion of ethanol. (S.C.E.E.B.)

2. 112 cm^3 of methane (measured at STP) was burned and the heat produced warmed 100 cm^3 of water from 15.0 °C to 25.5 °C. Calculate the heat of combustion of methane.

3. From the heats of combustion on page 202, calculate the heat of formation of (a) methanol, CH_3OH, (b) cyclohexane, C_6H_{12}.

4. When a mixture of aluminium powder and iron(III) oxide is ignited, iron is produced. The heat of formation of iron(III) oxide is – 827 kJ mol^{-1} of oxide; that of aluminium oxide is – 1676 kJ mol^{-1} of oxide.
(a) Write the equation for the action of aluminium on iron(III) oxide.
(b) Calculate the heat given out in the reduction of 1 mole of iron(III) oxide by aluminium. (S.C.E.E.B.)

5. $H_{2(g)} \longrightarrow 2H_{(g)}$ $\Delta H = +435$ kJ

$Br_{2(g)} \longrightarrow 2Br_{(g)}$ $\Delta H = +192$ kJ

$2HBr_{(g)} \longrightarrow 2H_{(g)} + 2\,Br_{(g)}$ $\Delta H = +728$ kJ

Calculate from these data
(a) the energy of the H–Br bond
(b) the heat of formation of hydrogen bromide from the gaseous elements. (S.C.E.E.B.)

6. The accepted value of the heat of combustion of methane is – 890 kJ mol^{-1}.

Given that $2\,CO + O_2 \longrightarrow 2\,CO_2$ $\Delta H = -568$ kJ

calculate the energy released in the partial combustion of 1 mole of methane to produce carbon monoxide and water. (S.C.E.E.B.)

7. Use the data shown to calculate
(a) the average C–Cl bond energy in tetrachloromethane, CCl_4
(b) the heat of formation of gaseous tetrachloromethane.

$CCl_{4(g)} \longrightarrow C_{(g)} + 4\,Cl_{(g)}$ $\Delta H = +1308$ kJ

$C_{(s)} \longrightarrow C_{(g)}$ $\Delta H = +715$ kJ

$Cl_{2(g)} \longrightarrow 2\,Cl_{(g)}$ $\Delta H = +243$ kJ

8.

bond	bond energy kJ mol^{-1}
N–H	391
N≡N	941
H–H	435

Use the values of bond energy in the table to calculate the heat of formation of 1 mole of ammonia from its elements. Show all your working and indicate clearly whether the reaction is exothermic or endothermic. (S.C.E.E.B.)

9. The reaction between hydrogen and chlorine is thought to involve the following stages:

$$Cl_{2(g)} \longrightarrow 2Cl\cdot_{(g)} \text{------------ (a)}$$
$$Cl\cdot_{(g)} + H_{2(g)} \longrightarrow HCl_{(g)} + H\cdot_{(g)} \text{-------- (b)}$$
$$H\cdot_{(g)} + Cl_{2(g)} \longrightarrow HCl_{(g)} + Cl\cdot_{(g)} \text{-------- (c)}$$

Use the mean bond dissociation energies on page 202 to calculate the heat of reaction for each of the processes (a), (b), and (c). Indicate by + or – sign which are endothermic and which exothermic. (S.C.E.E.B.)

10. Use the information on bond energies given on page 202 to calculate the heat of reaction for the complete hydrogenation of ethyne. (S.C.E.E.B.)

11. The heat of formation of cyclopropane is endothermic, energy value: 55 kJ mol^{-1}. The structural formula of cyclopropane, C_3H_6, is given below.

```
       H   H
        \ /
         C
        / \
  H    /   \    H
   \  C --- C  /
     /       \
    H         H
```

(a) Write the equation for the reaction referred to in the first sentence and alongside write the enthalpy change using the appropriate symbols.
(b) Use this enthalpy change and any other data you require from page 202 to calculate the average carbon-carbon bond energy in cyclopropane.
(c) Refer to the structural formula given and suggest why this value is considerably less than the value given on page 202 for a carbon-carbon single bond.
(d) Use the heat of formation given above and the heats of combustion of carbon and hydrogen given on page 202 to calculate the heat of combustion of cyclopropane.

Appendix – Experimental Measurement of Enthalpy Changes

In the following experiments, two important assumptions are made, namely
1. The density of a dilute aqueous solution is equal to that of water

$$1.00 \times 10^3 \text{ kg m}^{-3} \text{ or } 1.00 \text{ g cm}^{-3} \text{ at } 20°C$$

2. The specific heat capacity of a dilute aqueous solution is equal to that of water

$$4.19 \times 10^3 \text{ J kg}^{-1} \text{ K}^{-1} \text{ or } 4.19 \times 10^{-3} \text{ kJ g}^{-1} \text{ K}^{-1}.$$

Heat of neutralisation

Record the temperature of the acid (t_1) and alkali (t_2) before mixing. The average initial temperature is then $\frac{1}{2}(t_1 + t_2)$. Mix the solutions and record the highest temperature (t_3). Using the solutions below, Δt should be about 13.5 °C.

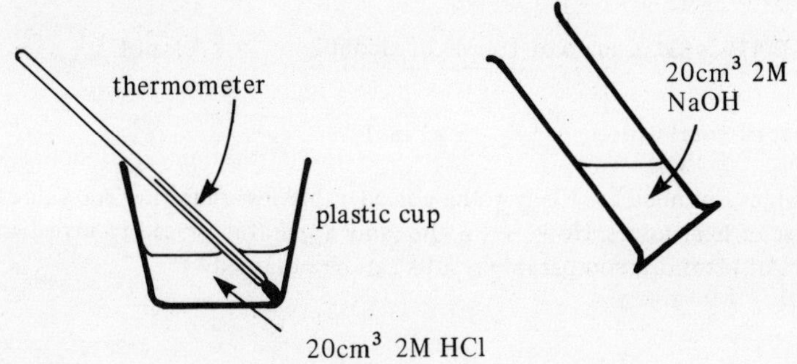

Fig. 4.5

temperature rise, $\Delta t = t_3 - \frac{1}{2}(t_1 + t_2)$ °C

total volume of solution $= 40$ cm^3

" mass " " $= 40$ g

quantity of heat released = (specific heat capacity of water) × (mass of water) × Δt

$= 4.19 \times 10^{-3} \times 40 \times \Delta t$ kJ

no. of moles of acid used = vol. in litres × molarity = 0.04

quantity of heat released per mole $= \dfrac{4.19 \times 10^{-3} \times 40 \times \Delta t \text{ kJ}}{0.04}$

$= 4.19 \times \Delta t$ kJ

i.e. heat of neutralisation, $\Delta H = -4.19 \times \Delta t$ kJ mol^{-1}

Heat of combustion

Weigh the burner before and after burning. Allow the burning alcohol to raise the temperature of the water by 10°C before extinguishing the flame. See fig. 4.6

mass of alcohol burnt = w g
mass of water heated in beaker = 100 g
rise in temperature = 10°C

quantity of heat released = (specific heat capacity of water) × (mass of water) × Δt

$$= 4.19 \times 10^{-3} \times 100 \times \Delta t \text{ kJ}$$

$$= 0.419 \times \Delta t \text{ kJ}$$

quantity of heat released per mole

$$= 0.419 \times \Delta t \times \frac{\text{mass of 1 mole of alcohol}}{w} = Y \text{ kJ mol}^{-1}$$

i.e. heat of combustion, $\Delta H = -Y \text{kJ mol}^{-1}$

The values obtained are likely to be considerably less than the true values because of heat losses. However, if the same apparatus is used to investigate a series of alcohols, comparable results can be obtained.

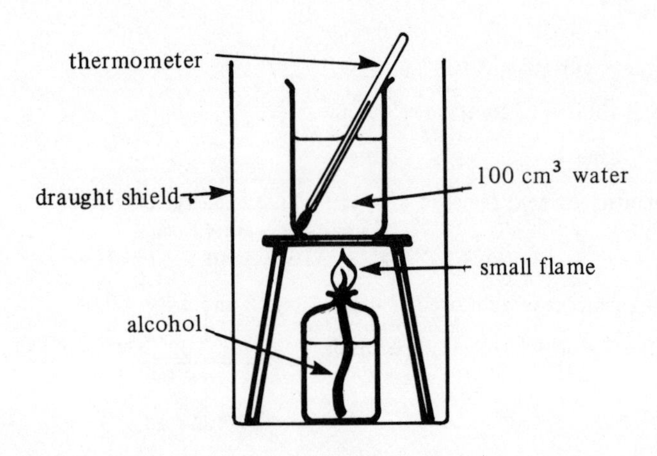

Fig. 4.6

Heat of solution

Weigh accurately about 1 g of the solute. Record the temperature of the water (t_1) and of the final solution (t_2).

mass of solute = w g

mass of water used = 50 g

temperature change, $\Delta t = t_2 - t_1$ °C

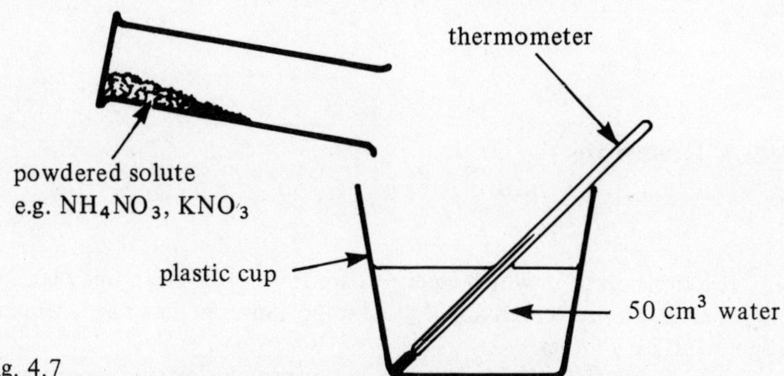

Fig. 4.7

quantity of heat change = (specific heat capacity of water) $\times$ (mass of water) $\times$ Δt

$$= 4.19 \times 10^{-3} \times 50 \times \Delta t \text{ kJ}$$

quantity of heat change per mole

$$= 4.19 \times 10^{-3} \times 50 \times \Delta t \times \frac{\text{mass of 1 mole of solute}}{w} \text{ kJ mol}^{-1}$$

$$= \text{Z kJ mol}^{-1}$$

i.e. heat of solution, ΔH = $-$ Z kJ mol^{-1}

Note: If reaction is endothermic, Δt will be negative and hence ΔH will be positive.

5 Competition for Electrons

Redox Reactions

Many reactions occur in which electrons are transferred from one reactant to another. Examples which are of great importance include displacement of one metal by another:
e.g. zinc + copper (II) sulphate soln. $\longrightarrow$ copper + zinc sulphate soln.
and displacement of hydrogen from an acid by a metal:
e.g. zinc + sulphuric acid $\longrightarrow$ hydrogen + zinc sulphate.
The first reaction above can be written as:

$$Zn_{(s)} + Cu^{2+}_{(aq)} + SO_4^{2-}{}_{(aq)} \longrightarrow Cu_{(s)} + Zn^{2+}_{(aq)} + SO_4^{2-}{}_{(aq)}$$

copper sulphate solution — zinc sulphate solution

i.e. Zn has changed to Zn^{2+} (loss of two electrons) and Cu^{2+} has changed to Cu (gain of two electrons) or two electrons have been transferred from Zn to Cu^{2+}

In the second reaction, Zn similarly loses two electrons to form Zn^{2+}, and the electrons are gained by the hydrogen ions of the acid solution.

$$Zn_{(s)} + 2H^{+}_{(aq)} + SO_4^{2-}{}_{(aq)} \longrightarrow H_{2\,(g)} + Zn^{2+}_{(aq)} + SO_4^{2-}{}_{(aq)}$$

sulphuric acid — zinc sulphate soln.

It is sometimes useful to separate the two parts of reactions of this type.

$Zn \longrightarrow Zn^{2+} + 2e$ (the loss of 2 electrons by Zn)

$Cu^{2+} + 2e \longrightarrow Cu$ (the gain of 2 electrons by Cu^{2+})

These 'half-reactions' written in this form are called ion-electron equations. They do not occur singly, but only in pairs involving both loss and gain of electrons.

The step involving loss of electrons is similar to the change of a metal when, for example, it forms its oxide. For example:

$Zn \longrightarrow ZnO$ containing Zn^{2+} ions.

Hence the loss of electrons has been given the general name of 'oxidation', although metals can lose electrons in many other reactions. Conversely, the gain of electrons is similar to the change of a metal oxide to a metal by reduction. For example:

CuO (containing Cu^{2+} ions) $\longrightarrow$ Cu.

Hence we commonly refer to gain of electrons as 'reduction'.

oxidation is the partial or total loss of electrons in a chemical reaction.
reduction is the partial or total gain of electrons in a chemical reaction.

Mnemonics which are useful in this connection are:-

Loss of **E**lectrons is **O**xidation $\longrightarrow$ L.E.O.
Reduction is **E**lectron **G**ain $\rightarrow$ R.E.G.

Oxidation of one reactant always involves reduction of another, therefore reactions involving these changes are often called **redox reactions**

A substance which brings about an oxidation, and is therefore reduced, is an **oxidising agent**. A substance which brings about a reduction, and is therefore oxidised, is a **reducing agent**.

Note: Not all chemical reactions involving ions are redox reactions Precipitation e.g.

$$Ag^+_{(aq)} + Cl^-_{(aq)} \longrightarrow AgCl_{(s)},$$

and neutralisation e.g.

$$H^+_{(aq)} + OH^-_{(aq)} \longrightarrow H_2O_{(l)}$$

do not involve electron transfer and are not redox reactions.

Demonstrating the transfer of electrons

The transfer of electrons in redox reactions is easily shown by setting up a 'cell'. e.g. in the case of the zinc/copper sulphate reaction, we can set up the two half-reactions

$$Zn_{(s)} \longrightarrow Zn^{2+}_{(aq)} + 2e$$

and $$Cu^{2+}_{(aq)} + 2e \longrightarrow Cu_{(s)}$$

in separate beakers linked by a voltmeter.

The introduction of a 'salt-bridge' completes a circuit and a reading is given on the meter. The reading indicates that electrons are flowing from zinc to copper i.e. the half-reactions are occurring as written and electrons *are* being transferred from zinc atoms to copper ions.

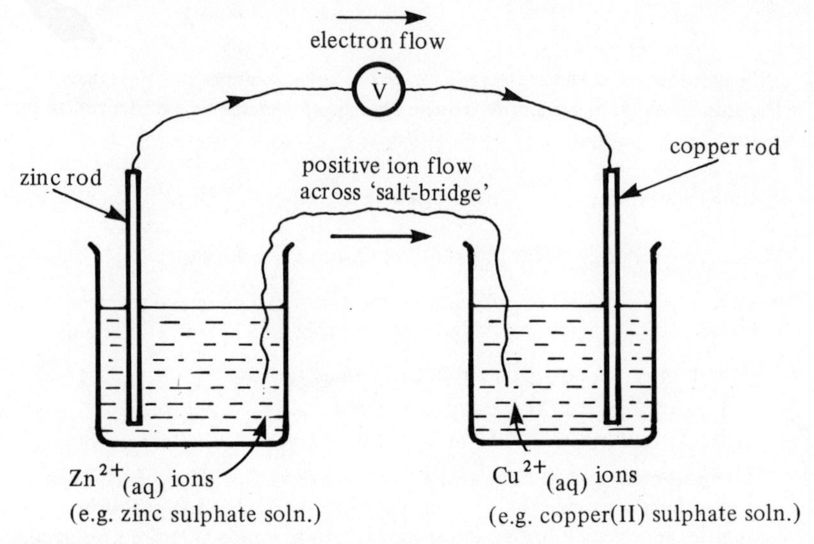

Fig. 5.1

A salt bridge is a means of conveying ions from one half of the cell to the other. It can simply be a filter paper soaked in a solution of an ionic material which does not react with the other solutions present, or an inverted U-tube containing the same solution.

In studies of the reactivity of metals, the different displacement abilities are often explained as the result of the ease with which metals form ions. Thus because magnesium forms an ion more readily than copper it will displace copper from copper sulphate, and copper will displace silver from silver nitrate for a similar reason.

In the zinc/copper cell of Fig. 5.1 a voltage was produced between the two metals. If magnesium/copper is used, the voltage is greater, and the electron flow is in the same direction – implying that the magnesium loses electrons more readily than zinc. If a silver/copper cell is set up, the voltage is small and the electron flow is reversed, hence silver loses electrons slightly *less* readily than copper. This provides a means of determining the ease of loss of or gain of electrons *compared with that of copper which we have used throughout.* The actual voltage is a measure of the overall energy change in the two half-cells *together*, and we cannot relate it directly to either copper or the other metal alone.

Measuring Standard Voltage

For a rigorous system, the conditions of measurement must be kept constant and internationally standard. Hence measurements are carried out at 25°C using molar solutions of reactants and, at a pressure of 1 atmosphere. Finally the reference material used is not copper, but hydrogen in a special device called a 'hydrogen electrode'. (See Fig. 5.2)

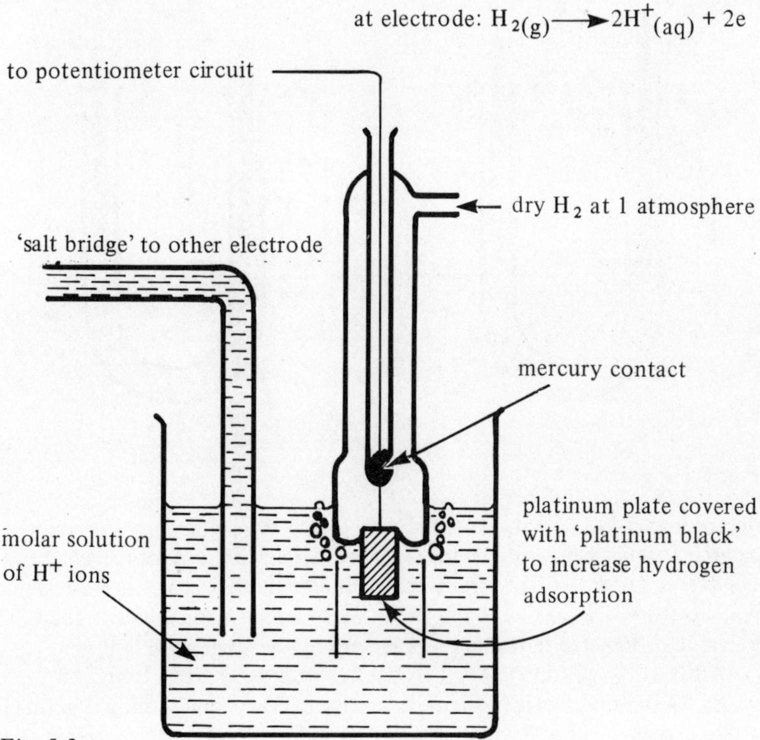

Fig. 5.2

A simple approximation to a hydrogen electrode can be made by electrolysing molar hydrochloric acid with carbon electrodes in a U-tube. (see Fig. 5.3). After a period of time, the solutions around the electrodes become saturated with gas, and the electrodes themselves adsorb some gas onto the surface. If the electrolysis is stopped, then the cell can be connected to a voltmeter when a value close to 1.3 volts is obtained. The cell gives the approximate E° value for the chlorine half-reaction and demonstrates the idea of reversibility.

The hydrogen electrode makes up one half of the overall cell, and it is paired with other half-cells, for example the zinc/zinc ion half-cell or the copper/copper ion half-cell. The voltage produced is of course dependent on the two half-cells, but since it always has a fixed component, namely the hydrogen electrode, it is sufficient to call this zero, and hence the cell's voltage is arbitrarily allocated to the half-cell paired with the hydrogen

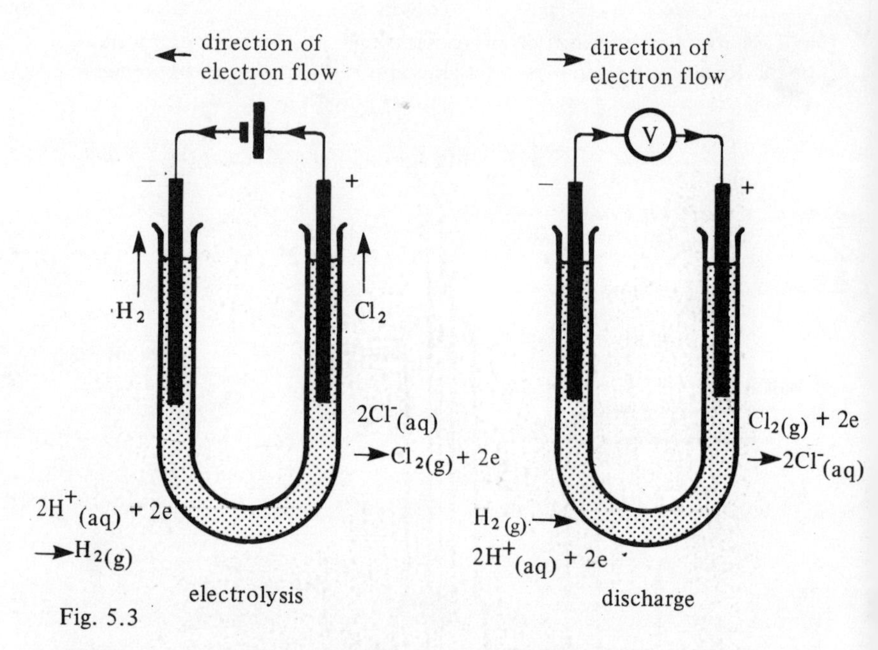

Fig. 5.3

electrode. i.e. the E° value of a half-reaction is the voltage it gives rise to when it forms a cell, under standard conditions, with a hydrogen electrode. The reactions occurring in these half-cells are always written as reductions (i.e. electrons being gained) and the voltage obtained for them is called the E° value or the **standard reduction electrode potential.**

A list of standard reduction electrode potentials will be found on page 203. It is important to remember that all the changes shown in this list are in fact reversible although it is the normal convention to represent them as *reductions.* The reversibility is easily illustrated in the case of a metal/metal ion half-reaction. For example:

$$Zn^{2+}_{(aq)} + 2e \longrightarrow Zn_{(s)}$$

This reaction occurs at the cathode during the electrolysis of an aqueous zinc salt solution. The reverse reaction occurs naturally when zinc metal is allowed to corrode forming various zinc salts, as happens to the casing of a dry cell.

Equations are written in the form:

electron acceptor or oxidising agent + electron(s) ⟶ electron donor or reducing agent

On Left Hand Side

the further *down* the table a reagent appears the more powerful an oxidising agent it is.

On Right Hand Side

the further *up* the table a reagent appears the more powerful a reducing agent it is.

OR

the further up the table a half-reaction appears the more likely it is to go in reverse,
and the lower down the table a half-reaction appears the more likely it is to go as written.

Using the E° Values

1. Predicting whether or not a reaction will occur

Since any of the reactions in the table of E° values is reversible, and the oxidising or reducing power of the reactions depends on the position in the table, we should be able to predict the course of a reaction, if any, between two substances in the table. Several examples follow.

Magnesium and lead nitrate solution. The important ion-electron equations are:

$$Mg^{2+}_{(aq)} + 2e \longrightarrow Mg_{(s)} \qquad E^\circ = -2.38\ V$$

$$Pb^{2+}_{(aq)} + 2e \longrightarrow Pb_{(s)} \qquad E^\circ = -0.13\ V$$

Magnesium is a better reducing agent than lead, as shown by its higher position in the E° table, and hence will be capable of reducing Pb^{2+} ions, i.e. magnesium will displace lead.

The actual reactions occurring will be therefore:

$$Mg_{(s)} \longrightarrow Mg^{2+}_{(aq)} + 2e \qquad E^\circ = +2.38\ V$$

$$Pb^{2+}_{(aq)} + 2e \longrightarrow Pb_{(s)} \qquad E^\circ = -0.13\ V$$

Since the first reaction is the reverse of the reaction in the E° table, the sign of the E° has been changed. The EMF of the cell reaction will be $+2.38 - 0.13 = +2.25\ V$

Chlorine and an iodide solution. The important ion-electron equations are:

$$I_{2(s)} + 2e \longrightarrow 2I^-_{(aq)} \qquad E^\circ = +0.54\ V$$

$$Cl_{2(g)} + 2e \longrightarrow 2Cl^-_{(aq)} \qquad E^\circ = +1.36\ V$$

As Chapter 9 shows, chlorine will displace iodine from an iodide solution, and so the half-reactions actually occurring are:

$$2I^-_{(aq)} \longrightarrow I_{2(s)} + 2e \qquad E^\circ = -0.54\ V$$

$$Cl_{2(g)} + 2e \longrightarrow 2Cl^-_{(aq)} \qquad E^\circ = +1.36\ V$$

The EMF of the cell reaction will be $-0.54 + 1.36 = +0.82$ V

It appears that reactions will occur if the EMF of the overall cell reaction is positive.

This can be tested if we consider:

Silver and copper sulphate solution. The ion-electron equations needed are:

$$Cu^{2+}_{(aq)} + 2e \longrightarrow Cu_{(s)} \qquad E^\circ = +0.34\ V$$

$$Ag^+_{(aq)} + e \longrightarrow Ag_{(s)} \qquad E^\circ = +0.80\ V$$

If a reaction were to occur, the half-reactions would be:

$$Cu^{2+}_{(aq)} + 2e \longrightarrow Cu_{(s)} \qquad E^\circ = +0.34\ V$$

$$Ag_{(s)} \longrightarrow Ag^+_{(aq)} + e \qquad E^\circ = -0.80\ V$$

The EMF of the overall cell reaction would be $+0.34 - 0.80\ V = -0.46\ V$

In fact silver will *not* displace copper from a solution of copper ions, and hence a negative value for the EMF of an overall cell reaction shows that the reaction is not possible.

Summary.

A redox reaction will probably occur if the *EMF of the overall cell reaction has a positive value.* When a reaction occurs the half-reaction lower in the E° table goes as written, reversing the upper half-reaction in the table.

2. Calculating cell voltages

If metals are paired in a solution of an electrolyte, the resulting cell will produce an EMF. The magnitude of the EMF can be predicted using E° values in a similar manner to that of the previous section. e.g. zinc/copper cell which is usually written as

$Zn_{(s)}/Zn^{2+}_{(aq)}/Cu^{2+}_{(aq)}/Cu_{(s)}$

i.e. zinc in a molar solution of a zinc salt, connected by a salt bridge to a molar solution of a copper salt containing a copper rod.

The half-reactions occurring are:

$$Zn_{(s)} \longrightarrow Zn^{2+}_{(aq)} + 2e \qquad E^\circ = +0.76\ V$$

$$Cu^{2+}_{(aq)} + 2e \longrightarrow Cu_{(s)} \qquad E^\circ = +0.34\ V$$

The overall cell reaction has an EMF of + 0.76 + 0.34 = 1.10 V

Similarly for magnesium and silver.

$Mg_{(s)}/Mg^{2+}_{(aq)}/Ag^{+}_{(aq)}/Ag_{(s)}$

the half-reactions occurring are:

$$Mg_{(s)} \longrightarrow Mg^{2+}_{(aq)} + 2e \qquad E^\circ = +2.38\ V$$

$$Ag^{+}_{(aq)} + e \longrightarrow Ag_{(s)} \qquad E^\circ = +0.80\ V$$

The EMF of the cell is + 2.38 + 0.80 = 3.18 V

Note: that in each case an ion-electron equation from the E° table has been reversed, and therefore the sign of its E° value has been changed. Once again, the reaction lower in the E° table goes as written, the upper one is reversed.

3. Predicting products of electrolysis

All electrolyses involve removal of electrons from ions at the anode i.e. oxidation, and gain of electrons by ions at the cathode i.e. reduction. *They are thus redox reactions.* The E° value table enables us to predict the product of an electrolysis where a choice of products is possible. For example, at the cathode the substance which is reduced most readily is to be expected i.e. the *cation lowest in the list* is likely to be discharged first.

Thus $\quad Cu^{2+}_{(aq)} + 2e \longrightarrow Cu_{(s)} \qquad E^\circ = +0.34\ V$

is likely to occur before

$$Sn^{2+}_{(aq)} + 2e \longrightarrow Sn_{(s)} \qquad E^\circ = -0.14\ V$$

At the anode, the ion which is most readily oxidised will appear, i.e. the *anion which is highest in the list* is likely to be discharged first.

Thus $2I^-_{(aq)} \longrightarrow I_{2\,(s)} + 2e \qquad E^\circ = -\,0.54\ V$

will occur before

$$2Cl^-_{(aq)} \longrightarrow Cl_{2\,(g)} + 2e \qquad E^\circ = -\,1.36\ V$$

If an aqueous fluoride solution is electrolysed, then

$$2H_2O_{(l)} \longrightarrow O_{2\,(g)} + 4H^+_{(aq)} + 4e \qquad E^\circ = -\,1.23\ V$$

will occur before

$$2F^-_{(aq)} \longrightarrow F_{2\,(g)} + 2e \qquad E^\circ = -\,2.87\ V$$

Note: in these last two examples where we considered the reverse of the reaction in the table, we have *changed the sign* of the E° value.

Limitations in the use of E° Values

The E° table is based on values obtained for substances used in their standard states, under standard conditions of temperature, pressure and concentration. If any of these conditions is varied, then predictions may be inaccurate.

1. Electrolysis

The products of electrolysis are often different from those predicted from E° values.

(a) Concentrations of *all* the species present are unlikely to be molar, and hence it is possible that a high concentration of one species may overcome its inferior discharge position in the E° table.

(b) The E° values for particular electrode reactions do not give any indication of *how quickly* a reaction will occur. They only indicate its relative feasibility. If two reactions are possible, then the faster reaction may occur, even if its E° value would tend to rule it out.

It is possible that either or both of the above factors are responsible for the formation of chlorine at the anode in the electrolysis of all but the most dilute chloride solutions. Possible anode reactions are:

$$4OH^-_{(aq)} \longrightarrow O_{2\,(g)} + 2H_2O_{(l)} + 4e \qquad E^\circ = -\,0.40\ V$$

$$2H_2O_{(l)} \longrightarrow O_{2\,(g)} + 4H^+_{(aq)} + 4e \qquad E^\circ = -\,1.23\ V$$

$$2Cl^-_{(aq)} \longrightarrow Cl_{2\,(g)} + 2e \qquad E^\circ = -\,1.36\ V$$

The discharge of OH^- ions is unlikely since in a neutral chloride solution the concentration of OH^- ions is much less than molar. The break up of water molecules is a slow process, and hence the discharge of chloride ions occurs, although this is the least favourable reaction from E° considerations. A further factor affecting ion discharge is the change of electrode material.

Change of electrode material. The products of a simple electrolysis can change if the electrode material is changed. For example in the electrolysis of sodium chloride solution using a mercury cathode, sodium is actually discharged at the cathode instead of hydrogen. The process is used industrially as mentioned in chapter 9. After a few minutes, the mercury surface is seen to evolve hydrogen at the R.H.S. The electrolysis has

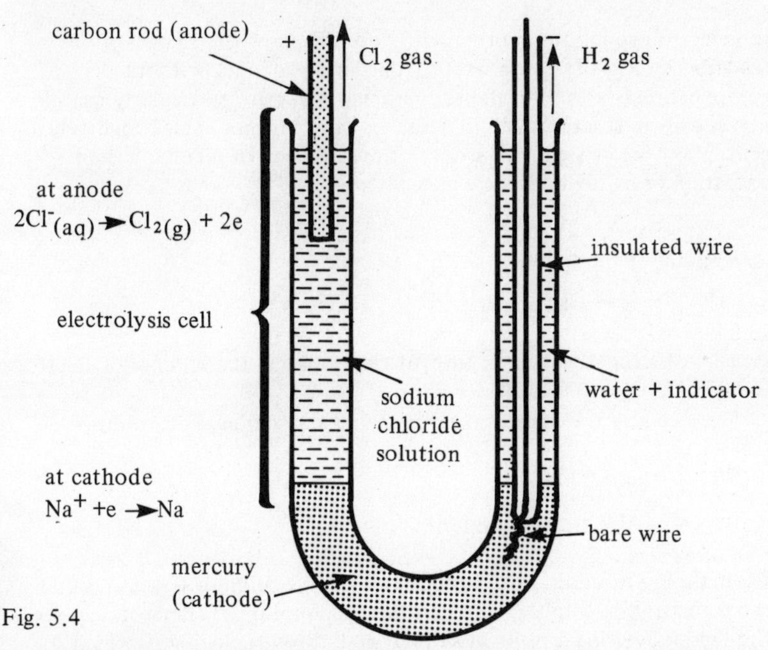

Fig. 5.4

produced sodium at the mercury cathode, which forms an amalgam i.e. an alloy of the metal with mercury. When the amalgam diffuses through to the water:

$$2Na/Hg + 2H_2O \longrightarrow 2Na^+_{(aq)} + 2OH^-_{(aq)} + H_{2\,(g)} + Hg$$

The formation of OH^- ions can be shown by adding indicator. If the cell is disconnected, and then reconnected to a voltmeter a potential difference of 2-3 volts is obtained as the reactions are reversed.

2. Cell potential differences

The cell potential differences for two half-cells calculated from E° values depend on the solutions used being molar. This is rarely the case in practice, and the potential difference can vary considerably. **Note**: the half-reactions alone are responsible for the cell potential difference. Any increase in the electrode size will not affect this potential difference, but the cell will be capable of delivering a greater current since more electrons can be transferred in the same time.

Writing Ion-Electron Equations

Ion-electron equations are frequently used to show the change of one reactant, by loss of or gain of electrons, into products, without the complication of other reactants appearing. They are particularly useful in electrolysis reactions and in describing the changes in electrochemical cells. Many are straightforward. e.g. the reduction of metal ions and oxidation of halide ions.

$$Cu^{2+}_{(aq)} + 2e \longrightarrow Cu_{(s)}$$

$$2Cl^-_{(aq)} \longrightarrow Cl_{2\,(g)} + 2e$$

In each case the direction of flow of electrons and the number of electrons is verifiable experimentally.

More complex are the changes involving oxyanions. For example

SO_3^{2-} – the sulphite ion.

$Cr_2O_7^{2-}$ – the dichromate ion.

Example 1. The addition of bromine water to a sulphite solution causes the formation of sulphate ions and the decoloration of the bromine (see the section revising aspects of sulphur and nitrogen chemistry which follows).

i.e. $$SO^{2-}_{3\,(aq)} \longrightarrow SO^{2-}_{4\,(aq)}$$

To gain oxygen atoms on one side of the equation, the appropriate number of water molecules is added and protons appear on the other side

i.e. $$SO^{2-}_{3\,(aq)} + H_2O_{(l)} \longrightarrow SO^{2-}_{4\,(aq)} + 2H^+_{(aq)}$$

Now the total charge on the left hand side is – 2 and on the right hand side is zero. Therefore we must *add* electrons to the right hand side.

i.e. $$SO_{3\,(aq)}^{2-} + H_2O_{(l)} \longrightarrow SO_{4\,(aq)}^{2-} + 2H^{+}_{(aq)} + 2e$$

Total charge on each side is now – 2.

Example 2. The addition of an aldehyde to a yellow-orange solution of potassium dichromate results in the formation of a green solution of Cr^{3+} ions (see chapter 16).

i.e. $$Cr_2O_{7\,(aq)}^{2-} \longrightarrow 2Cr^{3+}_{(aq)}$$

This time we have an excess of oxygen atoms to remove, and this step is accomplished by assuming combination with hydrogen ions (an acid solution is used).

i.e. $$Cr_2O_{7\,(aq)}^{2-} + 14H^{+}_{(aq)} \longrightarrow 2Cr^{3+}_{(aq)} + 7H_2O_{(l)}$$

Now the charges need balancing. On the left hand side, total charge is + 12, on the right hand side, total charge is + 6. Therefore we must *add* 6 electrons to the left hand side.

i.e. $$Cr_2O_{7\,(aq)}^{2-} + 14H^{+}_{(aq)} + 6e \longrightarrow 2Cr^{3+}_{(aq)} + 7H_2O_{(l)}$$

Total charge on each side is now + 6.

Redox Concepts concerned with the Sulphur and Nitrogen sections

In the 'O' Grade Course, various oxidising or reducing properties are stressed in the sulphur and nitrogen sections. It may be helpful to mention them again here.

1. Reducing properties of sulphur dioxide in solution

i.e. of a sulphite solution.
(a) Decoloration of bromine water.
(b) Decoloration of iodine solution.
(c) Reduction of Ag^{+} ions to metallic silver.
(d) Reduction of Fe^{3+} ions to Fe^{2+}

In each case the sulphite solution changes to a sulphate solution. The table of E° values gives:

			E°(V)
$SO_{4\,(aq)}^{2-} + 2H_{(aq)}^{+} + 2e$	$\longrightarrow$	$SO_{3\,(aq)}^{2-} + H_2O$	+ 0.17
$I_{2\,(s)} + 2e$	$\longrightarrow$	$2I_{(aq)}^{-}$	+ 0.54
$Fe_{(aq)}^{3+} + e$	$\longrightarrow$	$Fe_{(aq)}^{2+}$	+ 0.77
$Ag_{(aq)}^{+} + e$	$\longrightarrow$	$Ag_{(s)}$	+ 0.80
$Br_{2\,(l)} + 2e$	$\longrightarrow$	$2\,Br_{(aq)}^{-}$	+ 1.07

In each case our rule states that the lower reaction will go as written, reversing one above it. Hence each reaction could be predicted. In each case the flow of electrons is from the sulphite solution, as can be shown in these simple ways. (See Figs. 5.5 and 5.6)

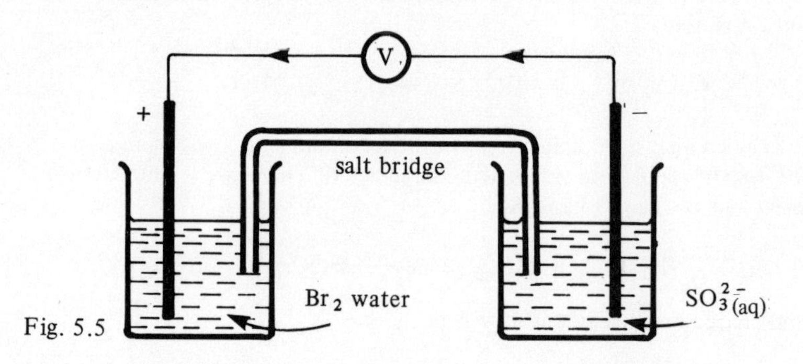

Fig. 5.5

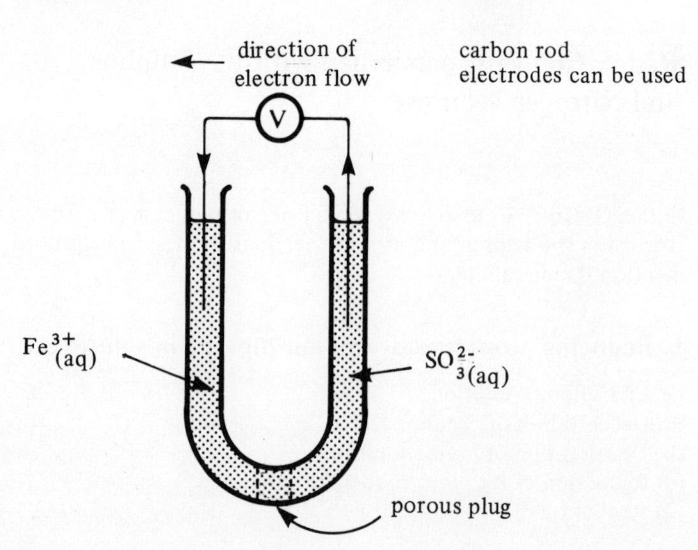

Fig. 5.6

2. Oxidising power of nitric acid

Unlike other acids, nitric acid generally does *not* yield hydrogen with metals. Instead it is reduced to form NO or N_2O_4 (usually visible as brown NO_2), the metal being oxidised in each case.

i.e. $Cu_{(s)} \rightarrow Cu^{2+}_{(aq)} + 2e$

$Mg_{(s)} \rightarrow Mg^{2+}_{(aq)} + 2e$

As will be seen from the E° values the NO_3^- ions are a stronger oxidising agent than the anions of most other acids.

When the acid is only moderately concentrated, the product is NO.

i.e. 1st Step $NO_3^-{}_{(aq)} \rightarrow NO_{(g)}$

2nd Step $NO_3^-{}_{(aq)} + 4H^+_{(aq)} \rightarrow NO_{(g)} + 2H_2O_{(l)}$

3rd Step $NO_3^-{}_{(aq)} + 4H^+_{(aq)} + 3e \rightarrow NO_{(g)} + 2H_2O_{(l)}$

derivation of ion electron equation

When the acid is concentrated, the product is N_2O_4.

1st Step $2NO_3^-{}_{(aq)} \rightarrow N_2O_4{}_{(g)}$

2nd Step $2NO_3^-{}_{(aq)} + 4H^+_{(aq)} \rightarrow N_2O_4{}_{(g)} + 2H_2O_{(l)}$

3rd Step $2NO_3^-{}_{(aq)} + 4H^+_{(aq)} + 2e \rightarrow N_2O_4{}_{(g)} + 2H_2O_{(l)}$

derivation of ion electron equation

The steps correspond to those in example 2 on page 53.

Examples for practice

1. Write ion-electron equations for:

(a) The conversion of iodate ions, IO_3^-, in acid solution to a solution of iodine.

(b) The change of PbO_2 to Pb^{2+} ions, and of Pb to Pb^{2+} ions, as occurs at the two electrodes during the discharge of a lead-acid accumulator.

2. Decide whether or not reactions are likely to occur in each of the following cases: (Assume all solutions are molar and at 25°C.)

(a) Acidified potassium manganate (VII) (permanganate) solution mixed with sodium chloride solution.

(b) Acidified potassium dichromate solution mixed with sodium chloride solution.

(c) Nitric acid mixed with potassium iodide solution.

3. What voltage would you expect to measure if cells were constructed from a silver plate immersed in molar silver(I) nitrate solution and

(a) mercury immersed in molar mercury (II) nitrate solution

(b) a tin plate immersed in molar tin (II) chloride solution?

4. Use the information on page 203.

The half-reaction equation for the reduction of the titanium(III) ion to titanium(II) ion is

$$Ti^{3+}_{(aq)} + e \longrightarrow Ti^{2+}_{(aq)} \qquad E^\circ = -0.37\ V$$

Which one of the following could be used as a reducing agent in this case?
(a) $Fe^{2+}_{(aq)}$ (b) $Zn_{(s)}$ (c) $MnO_{4\,(aq)}^{-}$

Write a half-reaction equation for the oxidation of the reagent you choose. (S.C.E.E.B.)

5. The half-reaction

$$O_{2\,(g)} + 2H^{+}_{(aq)} + 2e \longrightarrow H_2O_{2\,(l)}$$

has a standard reduction electrode potential of + 0.68 V.

Which ONE of the following reagents should be able to convert hydrogen peroxide (H_2O_2) into oxygen gas? (Use page 203)

$$Fe^{3+}_{(aq)};\ Fe^{2+}_{(aq)};\ Sn^{4+}_{(aq)};\ Sn^{2+}_{(aq)}$$

For the reagent you choose, write a half-reaction equation showing its reaction. (S.C.E.E,B.)

6. (You may find the data on page 203 useful in answering this question.)

The purple permanganate ion, $MnO_{4\,(aq)}^{-}$, reacts with the almost colourless manganese(II) ion, $Mn^{2+}_{(aq)}$, with the production of the brown solid, manganese(IV) oxide, MnO_2.

The half-reactions and the corresponding E° values are:

$$MnO_{2\,(s)} + 4H^{+}_{(aq)} + 2e \longrightarrow Mn^{2+}_{(aq)} + 2H_2O \qquad E^\circ = +1.23\ V$$

$$MnO_{4\,(aq)}^{-} + 4H^{+}_{(aq)} + 3e \longrightarrow MnO_{2\,(s)} + 2H_2O \qquad E^\circ = +1.67\ V$$

(a) A pupil uses the apparatus shown below to couple these reactions into a cell.

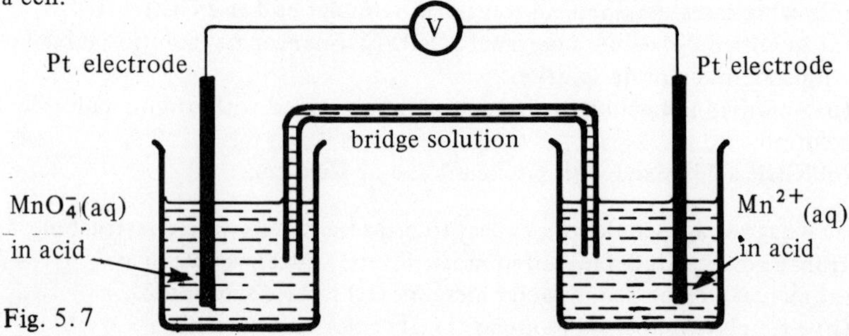

Fig. 5.7

(i) In which direction will electrons flow through the meter?
(ii) Which solution will act as oxidising agent?
(iii) What **two** properties would a substance require to make it a suitable 'bridge solution'?
(iv) What e.m.f. would the pupil calculate for the cell from the equations?
(v) Suggest a reason why he would not obtain this value on the voltmeter V.
(b) When a solution of manganese(II) sulphate is electrolysed between platinum electrodes a purple colour appears at the anode.
(i) What species causes the purple colour?
Give an ion-electron equation for its production from the $Mn^{2+}_{(aq)}$ ion.
(ii) Why does the purple colour appear at the anode and not at the cathode?
(iii) What ion might be discharged at the cathode? Give an ion-electron equation for the process.
(c) You are given four wires – magnesium, silver, tin, and zinc. They are similar in appearance, and you have to find out which is which. How would you try to do this if you had available in addition to the wires only a centre-zero galvanometer, and a beaker of dilute potassium nitrate solution? (S.C.E.E.B.)

7. In a laboratory experiment the following apparatus was set up:

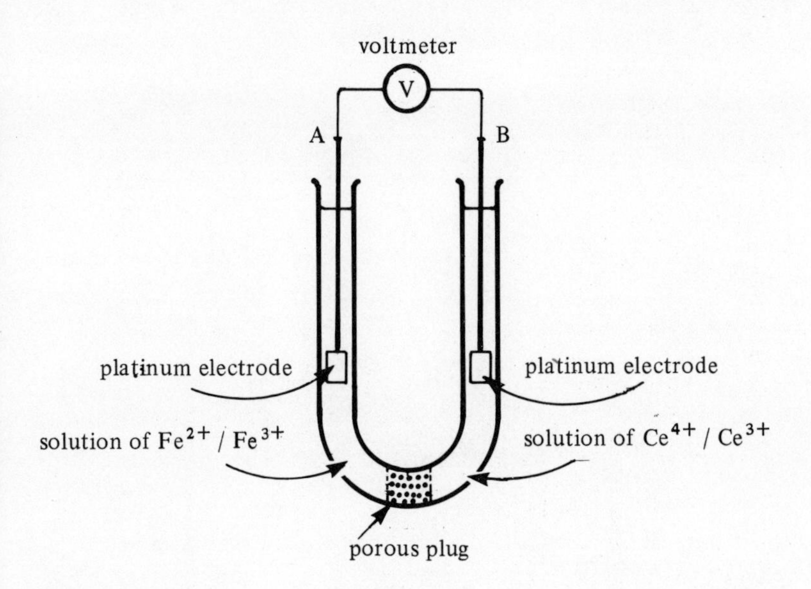

Fig. 5.8

The half reaction $Ce^{4+}_{(aq)} + e \longrightarrow Ce^{3+}_{(aq)}$ has a standard reduction electrode potential of + 1.44 V. Use this information and that on page 203 to answer the following questions:

(a) In the above cell, if all ionic concentrations are equal, in which direction will electrons flow in the external circuit?

(b) Assuming standard conditions, what reading will appear on the voltmeter?

(c) Write a balanced equation for the complete redox reaction occurring in the apparatus. (S.C.E.E.B.)

6 Elements and the Periodic Table

The Development of the Periodic Table

The Periodic Table will already be familiar, but more detailed study is now necessary. It is worthwhile seeing how the shape of the periodic table has developed. Nineteenth Century chemists had a bewildering mass of chemical information available to them. In order to try to simplify this information, and perhaps find some basic underlying principles, they devised various classifications of the elements. At this time the only apparently fundamental information available for atoms of different elements was the atomic weight, and this was made the basis of classification.

Various groupings were tried, and most were unsuccessful or derisively treated by a sceptical chemical 'Establishment'. However, in the Mid 19th Century Lothar Meyer began to accumulate data on the elements and investigate the variation with atomic weight. He discovered that many important quantities which could be measured accurately varied periodically, i.e. in a regular way, when plotted against atomic weight. Quantities in this category include melting point, boiling point, and atomic volume (the volume occupied by the atomic mass in grammes of the element in the solid state). A graph of this type (but using atomic number instead of atomic weight as the independent variable) shown in Fig. 6.1. It can be seen that alkali metals, halogens and noble gases lie at characteristic positions on the series of undulations produced.

The greatest step in the progress toward a periodic law was taken by Mendeleev in 1869. His basic statement was much the same as an earlier one of Newlands, that the elements fall into a repeating pattern of similar properties if arranged in order of increasing atomic weight. However, Mendeleev separated the list into vertical and horizontal sequences, groups and periods. More importantly he left blanks for as yet unknown elements, if it appeared that without these blanks dissimilar elements would be thrown together. He was able to leave blanks for elements discovered later – including gallium and germanium – and even to accurately predict their properties. In addition he was able to point to errors in atomic weight determinations, particularly of beryllium.

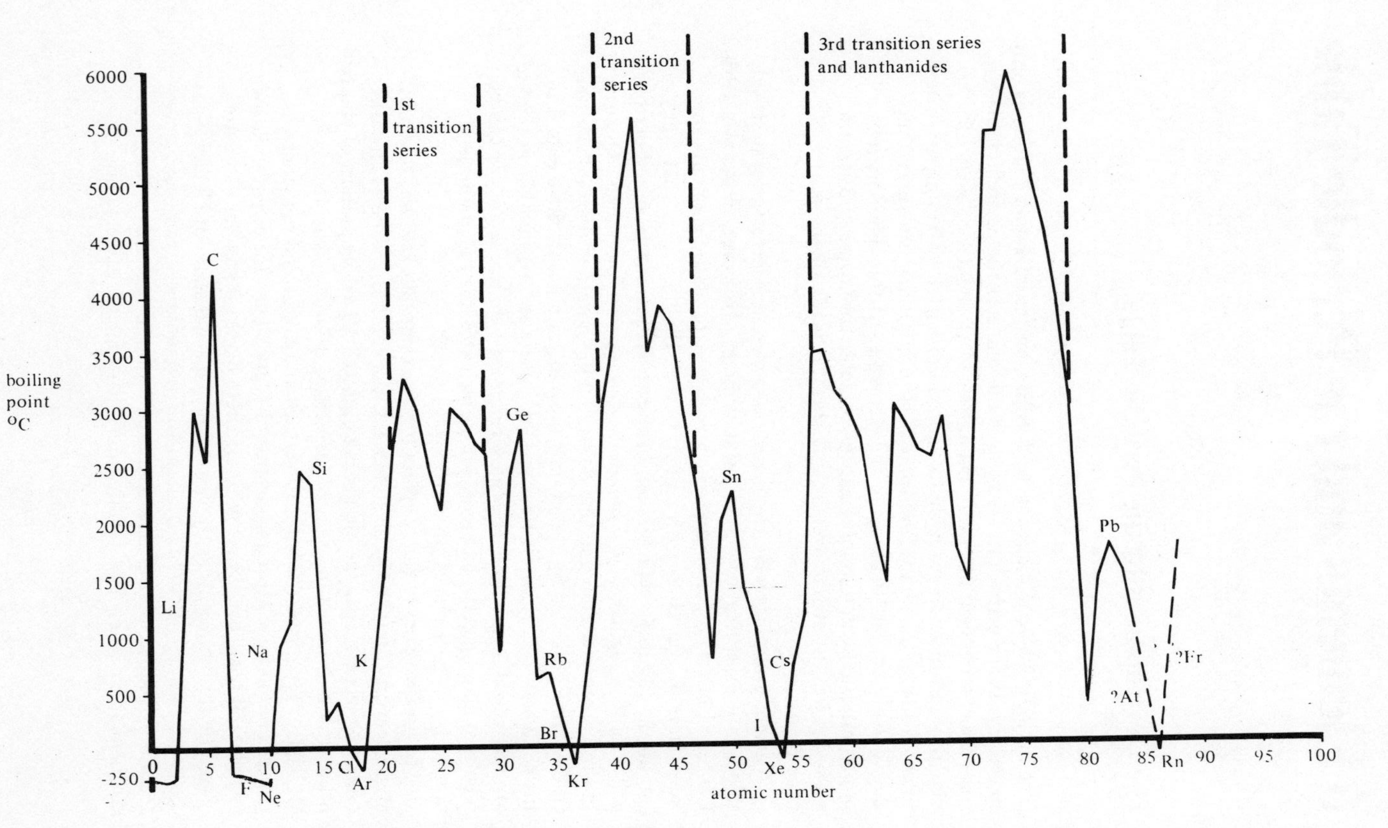

Fig. 6.1 variation of boiling point with atomic number

Despite the great advance made, and the ready acceptance of the periodic law, there were still many anomalies e.g. certain elements in the table were in reverse order of atomic weight and there was no easy way of placing the 'rare earths' – the elements lanthanum to lutetium.

The true significance of the periodic law became apparent with the discovery of 'atomic number' by Moseley in the early 20th Century. It then became apparent that the atoms of the elements were related in having an increasing number of protons and electrons which largely determined the chemical properties of the elements. Most of the anomalies of the periodic table were removed when classification was made by atomic number.

In the modern periodic table, the elements are arranged by increasing atomic number, each new horizontal row – period – commencing when a

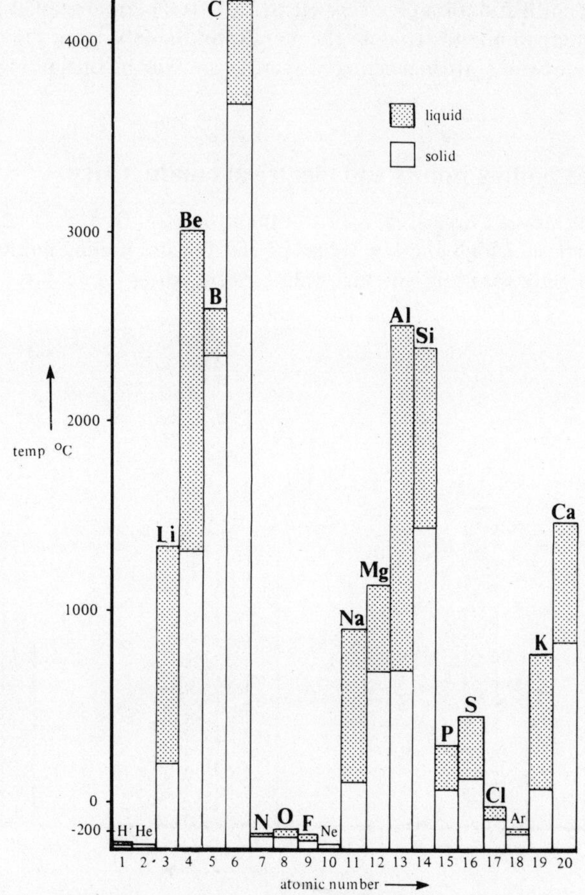

Fig. 6.2. melting and boiling points of elements 1–20

new layer of electrons starts to fill in the atom. Each main vertical column contains elements with the same number of electrons in the outermost layer. Since valency depends on the number of unpaired outermost electrons the main vertical groups contain elements with the same main valency e.g. group I the alkali metals of valency 1, group VII the halogens of valency 1 etc.

We shall be making use of the ideas of Lothar Meyer and Mendeleev to try and achieve a better understanding of the relationships between the first twenty elements. In any study of chemistry, a rough classification of the elements into metals and non-metals is introduced at an early stage. By consideration of, for example, metallic lustre, electrical conductivity, strength, melting point and boiling point about 75% of the 90 or so natural elements are found to be metallic, although it is often difficult to make a clear cut distinction for every element. Often periodic tables have a diagonal, stepped line to separate the metals, on the left, from the non-metals. Some of these distinguishing properties are worth considering again here.

Melting and boiling points and electrical conductivity.

In the Lothar Meyer Curve (Fig. 6.1) and the histogram (Fig. 6.2), the repeating pattern of high and low values of melting and boiling points can be seen. Generally speaking, the high values occur for elements just

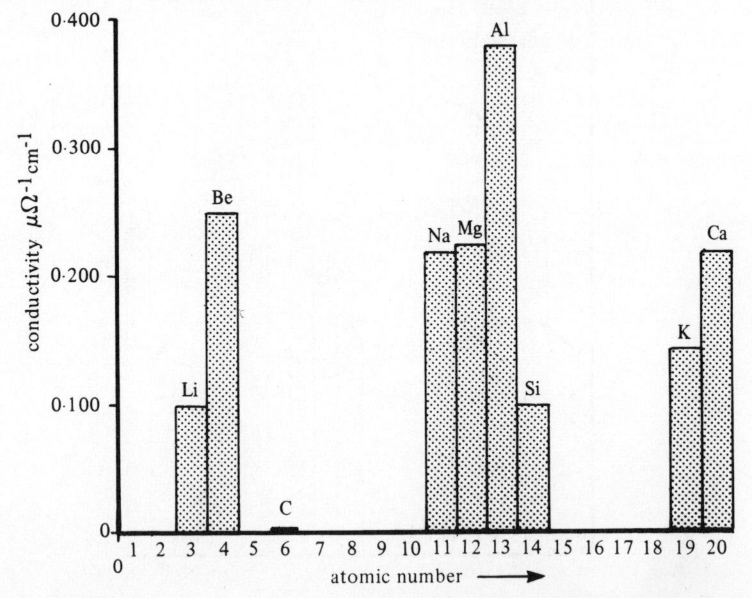

Fig. 6.3 variation of electrical conductivity with atomic number

starting a new layer of electrons, with a decline as the layer fills. High values indicate a large energy input is necessary to separate the particles of the elements sufficiently that they first become liquid and then vapour. In other words, the bonding between the particles of the elements on the left of the periodic table is stronger, or more extensive, than that between the particles of the right hand elements.

In Fig. 6.3 the electrical conductivity is seen to be high when a new layer of electrons is started, and to be virtually non existent for the elements whose outer layer is almost complete. If we assume that electrical conductivity depends on movement of electrons, it would seem that some of the electrons in the elements at the left of a row are freer than the electrons in elements at the right of the row.

The Bonding of the First 20 Elements

Group VIII elements – the noble gases

Each of these elements has a filled outer layer of electrons. Helium, neon and argon do not form compounds, and according to normal ionic and covalent bonding theory their atoms cannot join with each other. Accordingly the noble gases occur as single atoms. Since they will liquefy and solidify there must be some attraction between the atoms, and this is by means of **van der Waals'** bonding.

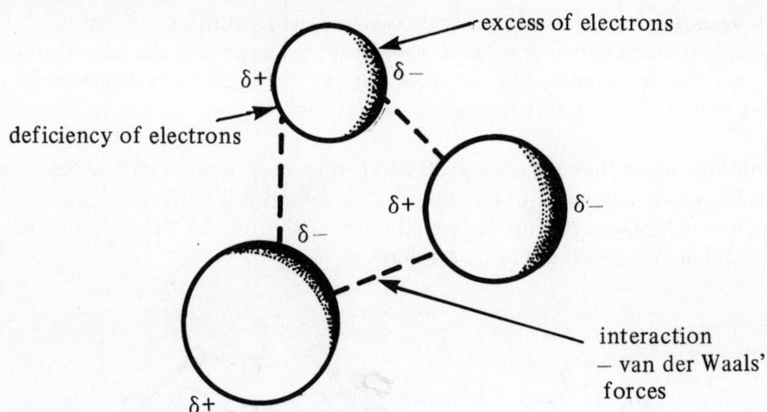

Fig 6.4

At any time the electrons in the atom may be distributed unevenly causing an 'instantaneous dipole' i.e. one side of the atom is slightly negative compared with the other side. These instantaneous dipoles can then attract each other to form a weak bond. The bond strength is of the order of 1-10 kJ mol^{-1} compared with 100 - 400 kJ mol^{-1} for covalent bonds.

In groups VII, VI, V and IV, the structures of the elements reflect covalent bond formation to achieve eight outer electrons.

Group VII elements

The halogens with one unpaired outer electron can form one covalent bond, and as a result diatomic molecules F_2, Cl_2, Br_2 and I_2 are formed. The molecules interact only weakly by the van der Waals' mechanism so that all the elements are volatile, and fluorine and chlorine are gaseous.

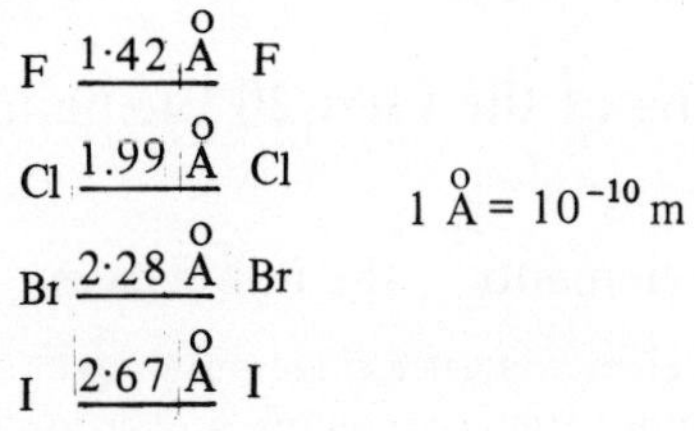

F — 1·42 Å — F

Cl — 1.99 Å — Cl

Br — 2·28 Å — Br

I — 2·67 Å — I

1 Å = 10^{-10} m

Fig. 6.5

Group VI elements

Oxygen. Each oxygen atom uses its two unpaired electrons to form two covalent bonds with one other oxygen atom (except when the rarer form ozone, O_3, is formed). The molecules interact by van der Waals' bonding, but since the interaction is weak, O_2 is gaseous.

Sulphur. When they have the ability to form two or more bonds, atoms can bond to more than one other atom. In the case of sulphur, closed, eight-membered, puckered rings are found in the orthorhombic variety, and zig-zag chains are found in plastic sulphur.

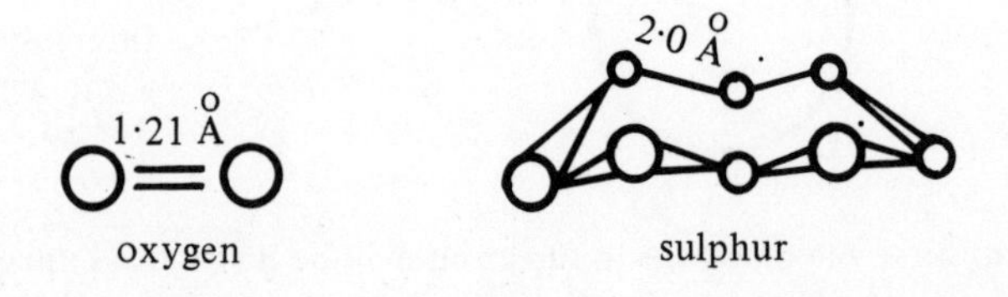

Fig. 6.6

Group V elements

Nitrogen. Nitrogen atoms form diatomic molecules with a triple bond, and only weak van der Waals' interaction.

Phosphorus. Phosphorus again makes use of single bonds to three other atoms to form tetrahedral P_4 molecules.

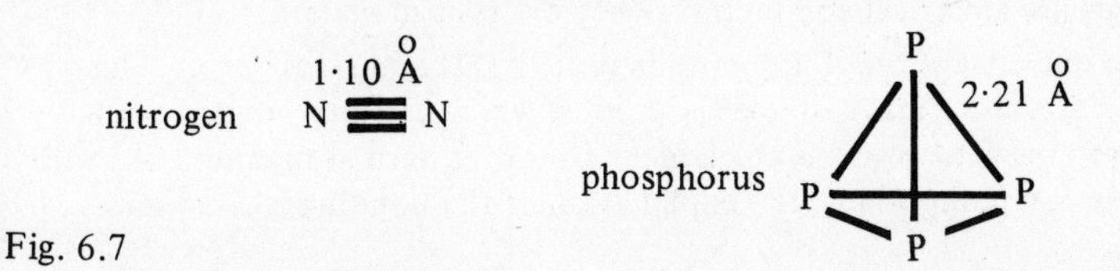

Fig. 6.7

Group IV elements

Neither carbon nor silicon form conventional multiple bonds between atoms, instead the 'standard' structure of the group is an infinite three-dimensional network, as in diamond and silicon. Each atom forms covalent bonds to four other atoms. The resultant structure is exceptionally hard and rigid. The other variety of carbon, graphite, has a structure based on three covalent bonds in one plane, and other reinforcing influences from delocalised electrons. It is illustrated in chapter 13.

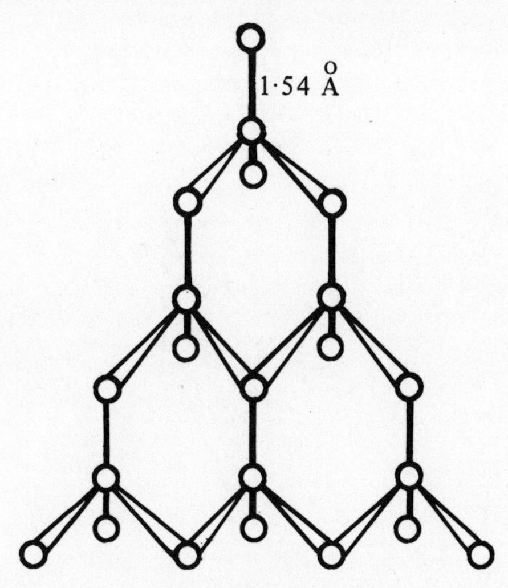

Fig. 6.8 diamond

Groups I, II and III.

There are insufficient electrons to allow the achievement of an octet by covalent bonding in these groups. Generally the elements contribute their outer electrons to a common 'pool' of delocalised electrons, which act as a binding medium for the resulting positive ions. The bonding is less directional than covalent bonding, and the metals are therefore malleable and ductile*. The electrons are capable of easy movements and hence the elements are electrical conductors. They are typical *metals.*

The one exception of any note in groups I, II and III is *boron.* This forms a structure made up of B_{12} groups, which are interbonded with other groups. The result is an element almost as hard as diamond. As with many boron compounds, the explanation of the bonding is complex.

* malleable – can be beaten or rolled into sheets.
ductilc – can be drawn into wire.

Examples for practice

These are to be found at the end of Chapter 7.

7 The Compounds of the 1st 20 Elements

The properties of all chemical compounds are governed by the type of bonding present, and therefore it is possible to infer the type of bonding within a compound by looking at its properties. The two extreme types of bonding are ionic and covalent. Both types aim at the achievement of a stable electron arrangement by the combining elements.

Ionic Bonding

1. Formation of a positive ion

The formation of a positive ion will be favoured by

(a) having to lose only a few electrons to achieve a stable external arrangement, i.e. the element will be in groups I, II or III;

(b) a relatively large atom, since the attraction by the nucleus for the outer electrons is reduced.

As the histogram of Fig. 7.1 shows, the atomic volume is relatively larger for groups I and II, where new layers of electrons have been started, than for succeeding groups, hence it is to be expected that elements of group I and II will lose electrons more readily. A measure of this ability is the **ionisation energy** of the element. This can be measured for the loss of successive electrons. Figures 7.2 and 7.3, show that for Groups I, II and III, the 1st, 2nd and 3rd ionisation energies respectively are relatively low, where only the outermost electrons are being removed, and that the removal of further electrons, from filled layers, requires much more energy. Thus, provided that this energy expenditure can be compensated by some exothermic step in compound formation, group I, II and III elements would be expected to form positive ions more readily than others.

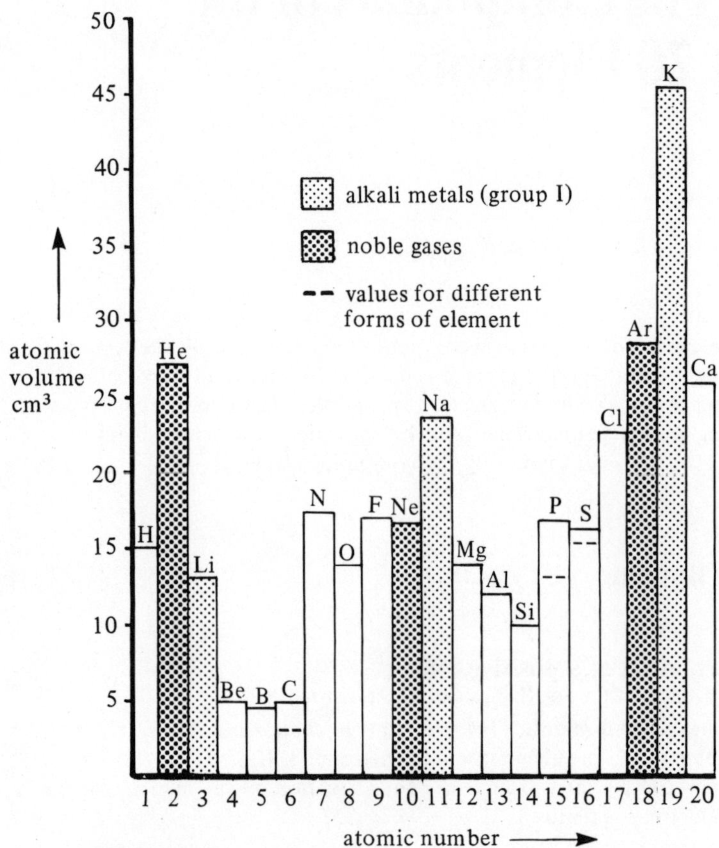

Fig. 7.1 variation of atomic volume with atomic no. (solid state)

2. Formation of a negative ion.

The formation of a negative ion will be assisted by:

(a) having to gain only a few electrons to achieve a stable electron arrangement i.e. the element will be in groups VII, VI and possibly V;

(b) relatively large nuclear charge – i.e. for the elements at the end of the horizontal period, the nuclear charge is greater than for those at the beginning, and electrons will be attracted into the outer layer more readily. The ease with which electrons are gained is measured by the **electron affinity** of the elements – usually an exothermic term.

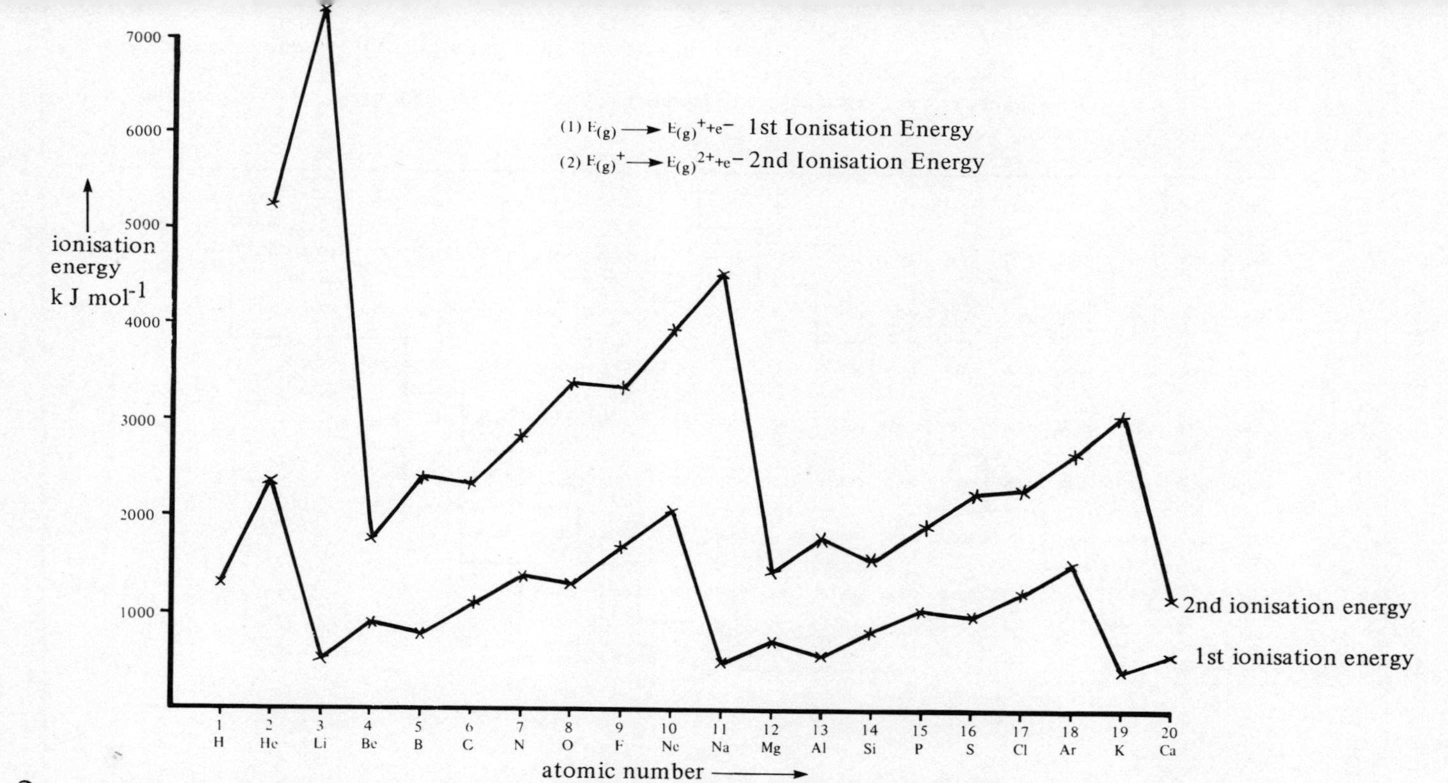

Fig. 7.2 1st 20 elements – variation of 1st and 2nd ionisation energies with atomic number

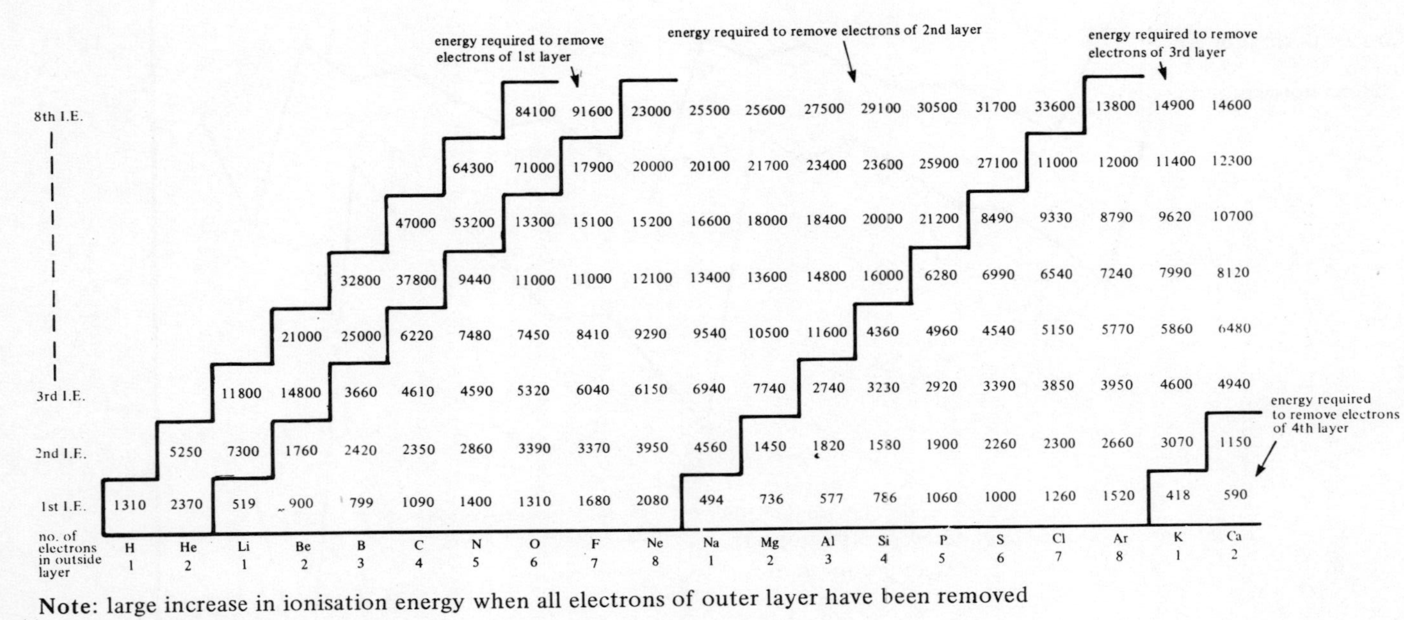

	H	He	Li	Be	B	C	N	O	F	Ne	Na	Mg	Al	Si	P	S	Cl	Ar	K	Ca
8th I.E.								84100	91600	23000	25500	25600	27500	29100	30500	31700	33600	13800	14900	14600
							64300	71000	17900	20000	20100	21700	23400	23600	25900	27100	11000	12000	11400	12300
						47000	53200	13300	15100	15200	16600	18000	18400	20000	21200	8490	9330	8790	9620	10700
					32800	37800	9440	11000	11000	12100	13400	13600	14800	16000	6280	6990	6540	7240	7990	8120
				21000	25000	6220	7480	7450	8410	9290	9540	10500	11600	4360	4960	4540	5150	5770	5860	6480
3rd I.E.			11800	14800	3660	4610	4590	5320	6040	6150	6940	7740	2740	3230	2920	3390	3850	3950	4600	4940
2nd I.E.		5250	7300	1760	2420	2350	2860	3390	3370	3950	4560	1450	1820	1580	1900	2260	2300	2660	3070	1150
1st I.E.	1310	2370	519	900	799	1090	1400	1310	1680	2080	494	736	577	786	1060	1000	1260	1520	418	590
no. of electrons in outside layer	1	2	1	2	3	4	5	6	7	8	1	2	3	4	5	6	7	8	1	2

Note: large increase in ionisation energy when all electrons of outer layer have been removed

Fig. 7.3 successive ionisation energies (kJ mol^{-1}) of 1st 20 elements

3. Formation of ionic bond

This is more likely to occur between elements at extremes of the periodic table, however, for most 'ionic' bonds, a degree of covalent character i.e. electron sharing will arise if the following criteria apply:

(a) the cation has a large charge and a small size, as for example Al^{3+}

(b) the anion has a large size e.g. I^-.

If these conditions are fulfilled, the cation will tend to distort the anion, drawing some of its electrons towards itself, and tending almost to the covalent situation.

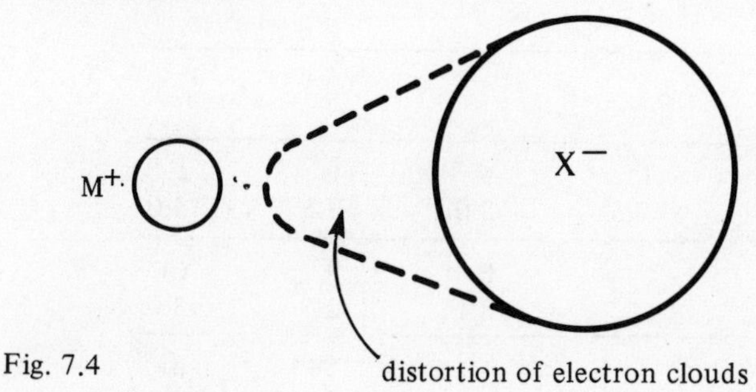

Fig. 7.4

Covalent Bonding

Covalent bonding will occur between elements which both require extra electrons to complete their outermost layer, in which case formation of a positive ion is impossible. The bonds form by overlap of electron orbitals which contain a single electron, and the overlapped orbitals with the shared pair of electrons constitute the covalent bond. The simplest picture of the way in which the bond holds the atoms together is to regard the shared electrons as providing attraction to the positive nuclei on either side.

A pure covalent bond i.e. one involving equal sharing of electrons can only occur when the atoms being joined are identical e.g. in chlorine, nitrogen or oxygen molecules. In any other situation, the dissimilar atoms will not share the electrons equally, and a **polar covalent bond** will result in a charge imbalance, represented by δ^+ and δ^-, on the molecule. For example

Fig. 7.5

The degree of polarity of the bond depends on the relative **electro-negativities** of the two elements being joined, i.e. the relative powers of the atoms in a molecule to attract electrons to themselves. This has been measured in a quantitative fashion, but it is generally agreed that too great an importance should not be attached to the values.

Electronegativity values are given in the data tables (pages 200 and 201), but a part of the table is reproduced here:

Table 7.1

H 2.1				
	C 2.5	N 3.0	O 3.5	F 4.0
		P 2.1	S 2.5	Cl 3.0
			Se 2.4	Br 2.8
			Te 2.1	I 2.5

Generally the electronegativity increases from left to right along a period since nuclear charge increases in the same direction, and the electro-negativity decreases down a group of the periodic table since the atomic size increases down the group.

The difference in electronegativity values for the atoms joined gives an indication of the relative degrees of polarity in covalent bonds. (See table 7.2)

Table 7.2

H – C 0.4	H – N 0.9	H – O 1.4	H – F 1.9
	H – P 0.0	H – S 0.4	H – Cl 0.9
		H – Se 0.3	H – Br 0.7
		H – Te 0.0	H – I 0.4

The polarity of the hydrogen-element bond is greatest for elements to the right of a period, or top of a group.

From the above discussion it will be seen that 'ionic' bonds commonly attain a degree of covalent character and *vice versa.* Hence it is often not strictly correct to speak of an 'ionic' or 'covalent' bond.

Properties of Compounds

Many of the points mentioned in the sections 'ionic bonding' and 'covalent bonding' are illustrated by the properties of the chlorides, oxides, and hydrides of the 1st 20 elements. The general properties of purely ionic and covalent compounds in table 7.3 should be borne in mind when reading the sections which follow. It is not intended that their information should be learned in detail. The general trends which are apparent are more important.

Table 7.3

ionic	covalent
lattice structure	discrete molecules
high melting point	low melting point
high boiling point	low boiling point
electrolytes i.e. conduct when molten or in solution, undergoing electrolysis	non-conducting
usually soluble in polar solvents and insoluble in non-polar solvents	generally more soluble in non-polar solvents
usually recovered unchanged from solution in polar solvents	often hydrolysed in water

1. Chlorides

The chlorides of the metallic elements can be made by direct combination, either by lowering the molten metal into a jar of chlorine, or by passing dry chlorine over the heated metal. The chlorides of sulphur and phosphorus can also be made by direct combination.

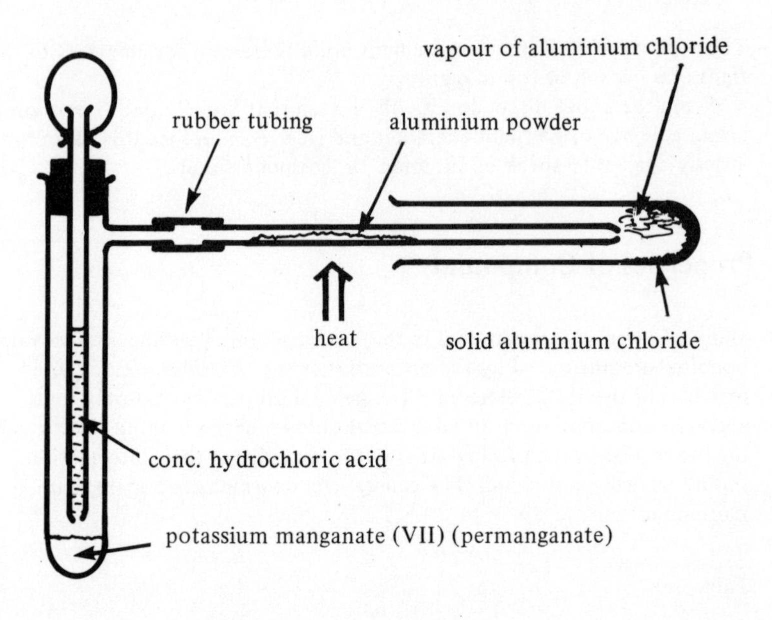

Fig. 7.6 preparation of anhydrous aluminium chloride.

The properties of the chlorides are summarised in Table 7.4. Hydrolysis of the covalent chlorides results in the formation of acidic solutions. For example:

$$SiCl_4 + 3H_2O \longrightarrow H_2SiO_3 + 4HCl$$

$$PCl_5 + H_2O \longrightarrow POCl_3 + 2HCl$$

2. Oxides

The oxides of most of the elements can be made by direct combination. Their properties are summarised in table 7.5. That of greatest importance is their behaviour with water. The oxides of group I, and also magnesium and calcium oxides produce alkaline solutions. For example:

$$Na_2O + H_2O \longrightarrow 2Na^+_{(aq)} + 2OH^-_{(aq)}$$

$$MgO + H_2O \longrightarrow Mg^{2+}_{(aq)} + 2OH^-_{(aq)}$$

Most of the oxides of groups V, VI,and VII, and carbon dioxide produce acidic solutions, e.g.

$$SO_2 + H_2O \longrightarrow H_2SO_3 \rightleftharpoons 2H^+_{(aq)} + SO_3^{2-}{}_{(aq)}$$

$$P_2O_5 + 3H_2O \longrightarrow 2H_3PO_4 \rightleftharpoons 6H^+_{(aq)} + 2PO_4^{3-}{}_{(aq)}$$

HCl −114 −85 72 forms hydrochloric acid δ+ δ− H—Cl molecular covalent gas			**KEY** empirical formula M.Pt. (°C) B.Pt.(°C) solubility (g/100g) structure and main bonding type			
LiCl 614 1350 83 sodium chloride lattice ionic solid	$BeCl_2$ 405 487 73 chain polymer covalent solid	BCl_3 −107 12 hydrolysed planar 120° molecule covalent gas	CCl_4 −23 77 sparingly soluble tetrahedral molecule covalent liquid	NCl_3 −27 71 sparingly soluble hydrolysed molecular unstable covalent liquid	OCl_2 −20 2 hydrolysed 115° bent molecule unstable covalent gas	ClF −154 −101 hydrolysed δ+ δ− Cl—F linear molecule covalent gas
NaCl 800 1465 36 sodium chloride lattice ionic solid	$MgCl_2$ 712 1418 55 layer stucture ionic solid	$AlCl_3$ 190 180* 46 partially hydrolysed dimeric covalent solid	$SiCl_4$ −68 57 hydrolysed tetrahedral molecule covalent liquid	PCl_3 −91 74 hydrolysed covalent liquid PCl_5 148 164 hydrolysed covalent solid	SCl_2 −80 59 hydrolysed δ+ S, δ− Cl 103° bent molecule covalent liquid	
KCl 770 1407 34 sodium chloride lattice ionic solid	$CaCl_2$ 772 — 74 distorted rutile structure ionic solid	*Sublimes		PCl_3: pyramidal molecule PCl_5: trigonal bypyramidal molecule (120°, 90°)		

IONIC. high melting point and boiling point stable in water → low melting point and boiling point **POLAR COVALENT** increasingly hydrolysed by water

Table 7.4

H_2O 0 100 amphoteric H–O–H 105° also H_2O_2 covalent liquids			KEY empirical formula M.Pt.(°C) B.Pt.(°C) acidity/alkalinity structure and main bonding type			
Li_2O > 1700 — strongly alkaline fluorite lattice ionic solid	BeO 2400 3900 amphoteric wurtzite lattice ionic/covalent solid	B_2O_3 577 — amphoteric glass, network structure covalent solid	CO / CO_2 −205 −190 / * −78 neutral / weakly acidic O = C = O covalent gases	N_2O / NO / NO_2 −103 / −164 / −9 −88 / −152 / 21 neutral / neutral / acidic dimeric N_2O_4 covalent gases/liquid		F_2O −224 −145 $\delta-$ F–O–F δ ($\delta+$ on O) 100° bent molecule covalent gas
Na_2O decomposes at high temp. strongly alkaline fluorite lattice ionic solid	MgO 2640 — weakly alkaline sodium chloride lattice ionic solid	Al_2O_3 2045 3000 amphoteric hexagonal ionic/covalent solid	SiO_2 1700 — amphoteric network structure of SiO_4 tetrahedra hexagonal covalent solid	P_2O_3 / P_2O_5 24 173 / 180* strongly acidic dimeric / dimeric covalent solids	SO_2 / SO_3 −73 −10 / 17 40 weakly acidic / strongly acidic $\delta-$ O–S–O $\delta-$ ($\delta+$ on S) 122° / trimeric covalent gas / covalent solid	Cl_2O −20 2 Cl–O–Cl ($\delta+$ on O) 115° also ClO_2, Cl_2O_7 covalent gas/liquids
K_2O decomposes at high temp. strongly alkaline fluorite lattice ionic solid	CaO 2700 — moderately alkaline sodium chloride lattice ionic solid	*sublimes				

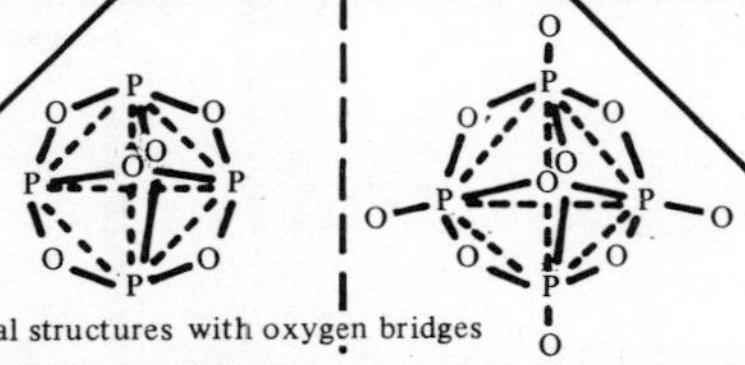

tetrahedral structures with oxygen bridges

IONIC alkaline ——— amphoteric ———→ **POLAR COVALENT** acidic

Table 7.5 Oxides

Some of the oxides e.g. N_2O and CO undergo no change with water (neutral oxides), and a fourth group, including the oxides of beryllium, aluminium and silicon, although not soluble in water alone, behave in an **amphoteric** fashion. Amphotericity is the ability of a substance to behave as an acid, in reaction with a base, or as a base, in reaction with an acid.

e.g. As a base:

$$Al_2O_3 + 3H_2SO_4 \longrightarrow Al_2(SO_4)_3 + 3H_2O \qquad \text{(Aluminium sulphate)}$$

As an acid:

$$Al_2O_3 + 2NaOH \longrightarrow 2NaAlO_2 + H_2O \qquad \text{(Sodium aluminate)}$$

These reactions are often difficult to perform with commercial aluminium oxide, but can be carried out with hydrated aluminium oxide, or aluminium hydroxide, precipitated from aluminium sulphate solution by sodium hydroxide solution. The precipitate can be filtered off and will dissolve in dilute acid or in excess sodium hydroxide solution.

3. Hydrides

The hydrides of groups I and II can be made by direct combination of the element with hydrogen, as can the hydrides of oxygen, sulphur, nitrogen and the halogens, although for H_2S and the hydrogen halides the normal method of preparation is to displace them from metal sulphides or halides using a strong mineral acid. Silane and phosphine are rather unusual in their preparation and are noteworthy for apparently being spontaneously inflammable.

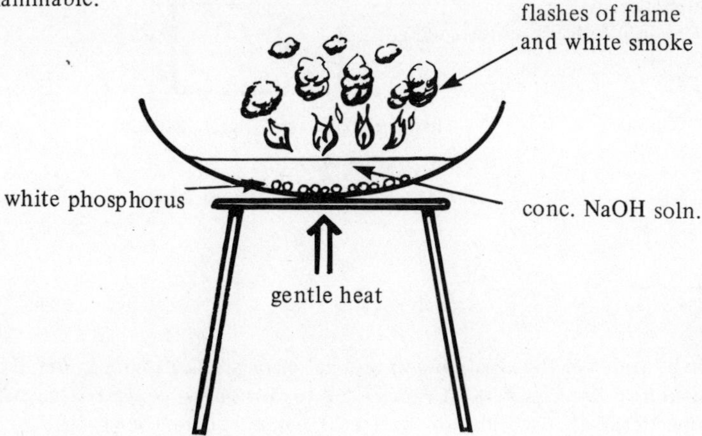

Fig. 7.7 preparation of phosphine PH_3

$$3NaOH + 4P + 3H_2O \longrightarrow PH_3 + 3NaH_2PO_2$$

The phosphine itself is inflammable, but the gas contains P_2H_4 which is spontaneously inflammable and so causes ignition on contact with air.

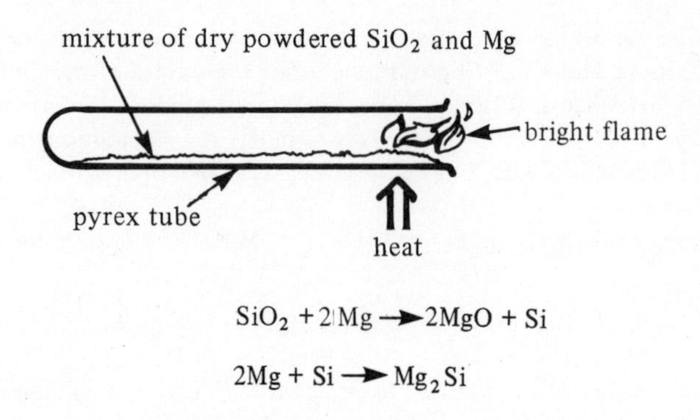

Fig. 7.8 and Fig. 7.9 Preparation of Silane SiH_4

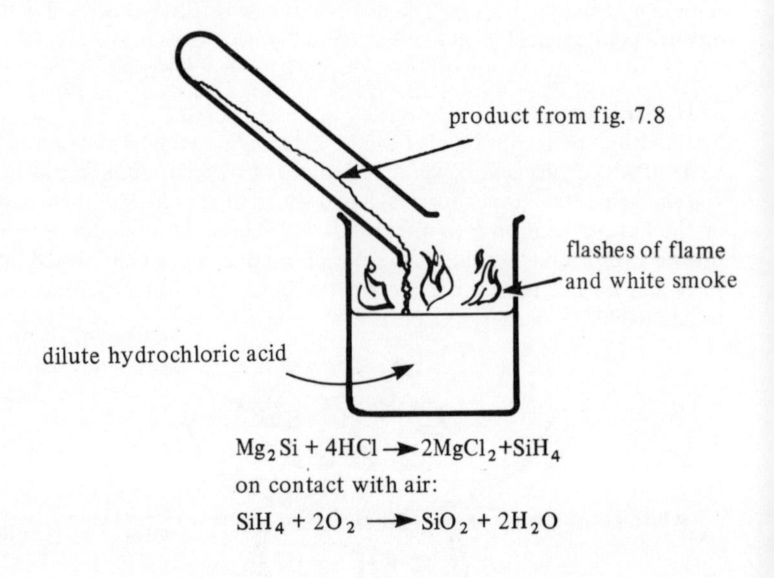

Fig. 7.9

The hydrides of the alkali metals and halogens are dealt with in detail in the appropriate chapters. It is intended to give only a general survey of the properties of the hydrides of the 1st 20 elements here. (See table 7.6)

The bonding of the boron hydrides, as with elemental boron, raises problems in that there appear to be too few electrons to allow normal covalent bonding. A satisfactory explanation has been made by the use of Molecular Orbital Theory, a more general theory than the simple theories of bonding which have proved sufficient for our purposes until now.

KEY: empirical formula M.Pt.(°C) B.Pt.(°C) solution formed with water structure and main bonding type						
LiH 680 alkaline, H_2 evolved sodium chloride lattice ionic solid	BeH_2 preparation difficult, little known of properties	BH_3 and others −169 −88 H_3BO_3 formed and H_2 evolved (diagram: H, B, H, B, H, H, H, H) dimeric	CH_4 and others −182 −160 neutral (diagram: C, H, H, H, H, 108°) tetrahedral covalent	NH_3 also N_2H_4 −78 −33 alkaline (diagram: N δ−, H δ+, H δ+, H δ+) pyramidal covalent	H_2O also H_2O_2 O 100 neutral (diagram: O δ−, H δ+, H δ+, 105°) covalent molecule	HF −83 19 acidic δ+ δ− H—F linear covalent
NaH 800 (dec.) alkaline, H_2 evolved sodium chloride lattice ionic solid	MgH_2 alkaline, H_2 evolved rutile lattice ionic/covalent solid	AlH_3 $(AlH_3)n$ structure uncertain	SiH_4 and others −133 −14 neutral (diagram: Si, H, H, H, H, 108°) tetrahedral covalent	PH_3 also P_2H_4 −133 −88 weakly alkaline (diagram: P, H, H, H) pyramidal covalent	H_2S −83 −62 weakly acidic (diagram: S, H, H, 92°) covalent molecule	HCl −114 −85 strongly acidic δ+ δ− H—Cl linear covalent
KH decomposes alkaline, H_2 evolved sodium chloride lattice ionic solid	CaH_2 816 alkaline, H_2 evolved orthorhombic lattice ionic solid					

IONIC ⟶ POLAR COVALENT

Table 7.6 Hydrides

Anomalous Physical Properties of Some Hydrides

1. Melting and Boiling Points

The accompanying graphs, Figs. 7.10 and 7.11, show for the group IV hydrides the expected increase in melting point and boiling point with formula mass. However, in the other three groups, the values of m.pt. and b.pt. for NH_3, H_2O and HF (and HCl to some extent) are higher than would be expected for their formula mass.

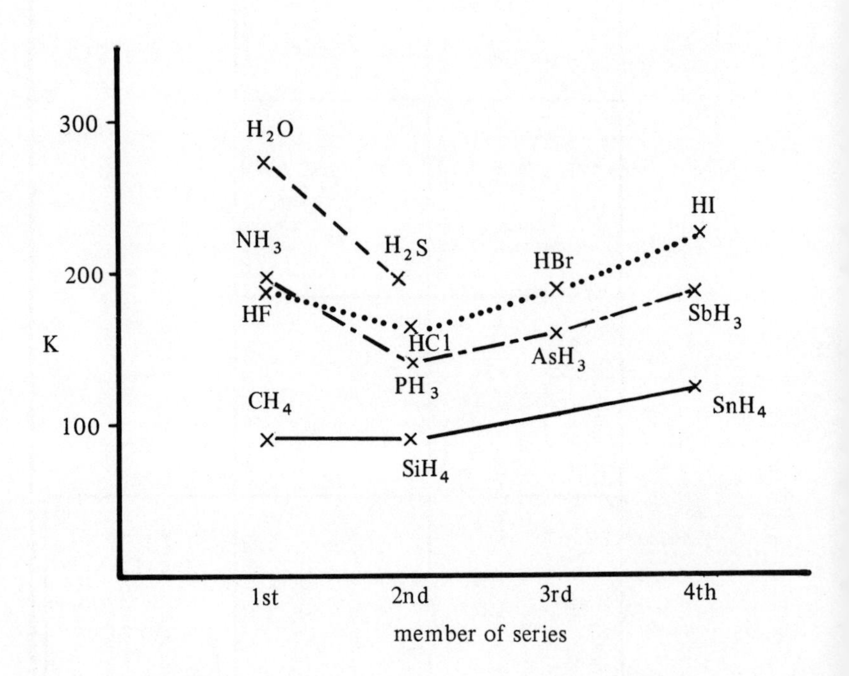

Fig. 7.10 melting points in K

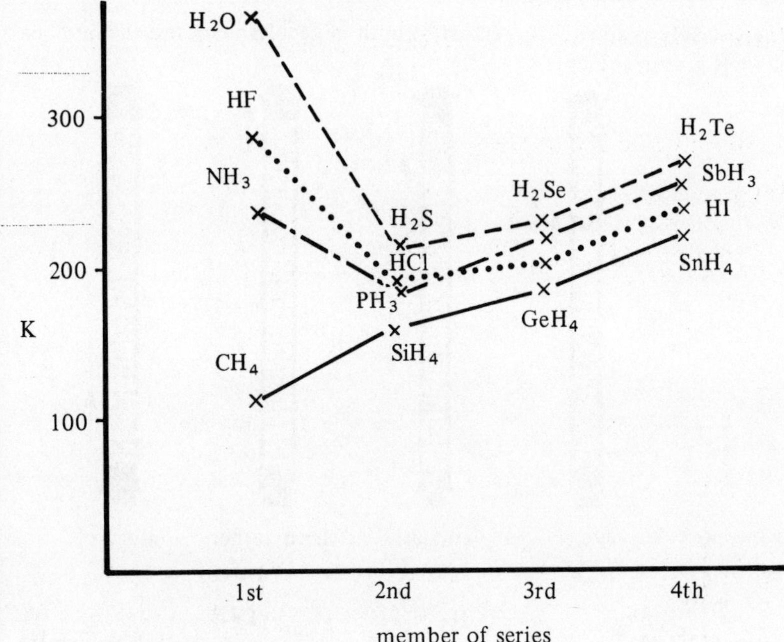

Fig. 7.11 boiling points in K

2. Latent Heats of Fusion and Evaporation

Table 7.7

	ΔH_{Fus} (kJ mol^{-1})	ΔH_{Evap} (kJ mol^{-1})
H_2O	+ 6.02	+ 41.04
H_2S	+ 2.47	+ 18.64

The latent heats for the changes solid to liquid and liquid to vapour are higher for the substance of lower molecular mass, contrary to the expected situation.

3. Surface Tension

Water has a far higher surface tension than other liquids of similar molecular mass, hence its ability to support dense objects such as carefully placed needles etc.

4. Viscosity

Compared with liquids of higher molecular mass, the viscosity of water is surprisingly high. This can be shown by simultaneously inverting tubes containing various liquids with an air bubble trapped at the top.

The viscosity is apparently related to both molecular mass and the number of –OH groups present.

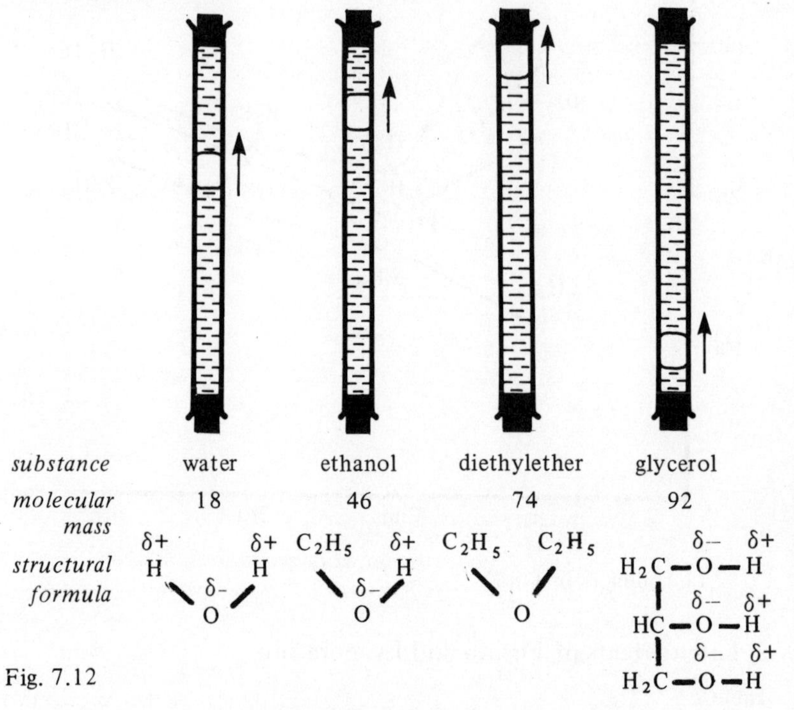

Fig. 7.12

These anomalous properties would appear to indicate stronger bonding between the molecules than the expected van der Waals' bonding (i.e. interactions due to a temporary imbalance of electrical charges). The compounds showing these anomalous properties all contain bonds which are distinctly polar i.e. O–H, N–H, F–H (and Cl–H) as shown by the difference in electronegativities in table 7.2. These molecules can therefore interact in this fashion:

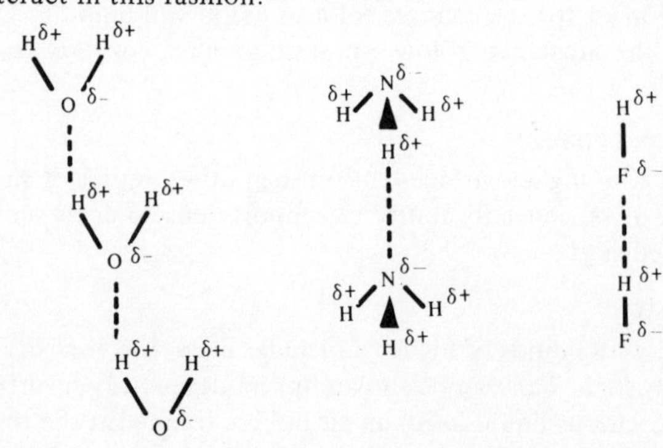

Fig. 7.13

Hydrogen Bonding. This weak interaction is called **hydrogen bonding** since it occurs only for compounds containing a strongly electronegative element linked to hydrogen. The pull of electrons away from the hydrogen results in a positive charge located on a small atom, and hence a high positive charge density capable of interacting with the negative ends of other molecules.

Hydrogen bonding is slightly stronger than van der Waals' bonding, about 30 kJ mol^{-1} compared with 4 kJ mol^{-1}, but weaker than covalent bonds which are of the order of 450 kJ mol^{-1}.

An important consequence of hydrogen bonding is the unusual way in which water freezes. As with all liquids, water contracts on cooling, but

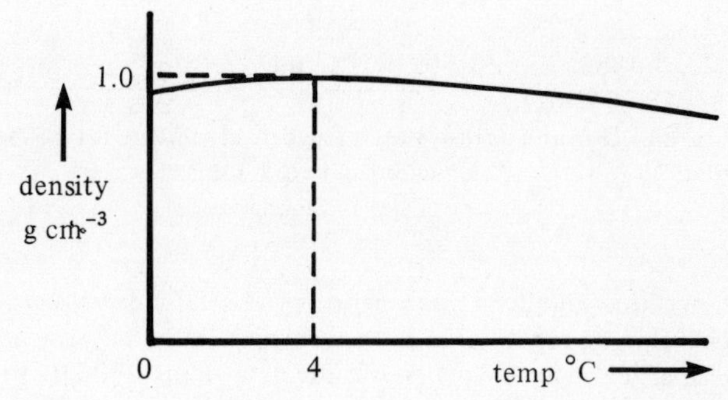

Fig. 7.14 (not to scale)

when it reaches 4°C it begins to expand again, and therefore at the freezing point is less dense than water which is about to freeze. The reason for this is the ordering of molecules into an open lattice as the hydrogen bonds are able to overcome the decreasing thermal motion of the molecules. As a result ice floats in water, seas freeze from the top downwards, allowing fish to survive in unfrozen water beneath (and allowing life to evolve in the sea without interruption) and of course pipes burst when water freezes inside them. In a biochemical context, hydrogen bonds are responsible for the binding together of the two helices of D.N.A.

Examples for practice

1.

substance	melting point (°C)	boiling point (°C)	electrical conductivity of solid	electrical conductivity of melt
A	92	190	nil	nil
B	1050	2500	good	good
C	773	1407	nil	good
D	1883	2503	nil	nil

Place A, B, C and D in the appropriate categories from the following:
metallic solid — covalent network solid
ionic solid — covalent molecular solid (S.C.E.E.B.)

2. As we move horizontally across a period or vertically down a group in the Periodic Table we can observe gradual changes in the physical and chemical properties of the elements. Use the data on pp. 200-201 to help you answer the following questions.
(a) Where, between elements of atomic number 1 and 12 are the most abrupt changes in the chemical properties?
(b) What two factors cause the tendency for the first ionisation energy to increase as we move across a period?
(c) Within the period sodium to argon write formulae for
  (i) the compound with the highest degree of ionic bonding.
  (ii) the hydride with the highest degree of covalent bonding.
(d) By means of extended structural formulae show why the group 4 elements carbon and silicon have dioxides which are totally different in character.
(e) At ordinary temperatures, five distinct structures exist among the elements. These are
  closely packed molecules,
  giant covalent structure,
  giant ionic structure with interstitial electrons,
  atomic gas,
  molecular gas.
Which of these best describe the normal state of sulphur, silicon, argon and aluminium?
(f) Among simple compounds, the oxides probably show the most regular trends as we move across a period. In what sense can aluminium oxide be

said to be intermediate in properties between the oxides of sodium and sulphur?

(g) Mendeleev was able to predict many properties of elements yet to be discovered. Predict the physical appearance and estimate the melting points of:

(i) francium.

(ii) astatine. (S.C.E.E.B.)

3.

(a) Of the elements of atomic number 3-10 choose one in each case which has a structure you would classify as

(i) metallic.

(ii) covalent, 'network' type.

(iii) covalent, 'discrete molecules.'

(b) Use the examples you have chosen for each of the types (i) – (iii) and the relevant data on pp. 200-201 to support the following statement:

A noteworthy difference between metallic and non-metallic elements (whether these are of 'network' or 'discrete molecule' type) is that the metals can exist as liquids over a wider range of temperature.

(c) State whether you would expect each of the following to conduct electricity appreciably when connected to a low voltage source.

(i) solid rubidium chloride.

(ii) liquid gallium (element 31).

(iii) liquid nitrogen.

Give your reason briefly in terms of the type of bonding present.

(d)

compound	formula	molecular weight	boiling point
ethane	CH_3CH_3	30	−89°C
methanol	CH_3OH	32	64°C
hydrazine	NH_2NH_2	32	113°C
silane	SiH_4	32	−112°C

(i) From the information given, which of the compounds in the table contain hydrogen-bonding in the liquid state?

(ii) Why does hydrogen-bonding affect the boiling point of a substance?

(ii) In the table we have compared substances of similar molecular weight. Why is molecular weight significant in this case?

(iv) State two other ways in which the presence of hydrogen-bonding could affect the physical properties of a substance. (S.C.E.E.B.)

4. TITANIUM

Occurrence and Extraction.

Titanium occurs in the form of the mineral rutile, TiO_2. Rutile can be converted to $TiCl_4$, a colourless liquid which boils at 136°C and fumes in moist air. The $TiCl_4$ is heated with magnesium, and the mixture resulting from this reaction is washed with very dilute acid, to leave titanium.

Properties

Titanium is resistant to corrosion by acid and by sea water, but will react if heated with fused alkali, to give substances called titanates, e.g. K_2TiO_3, potassium titanate.

Uses

Titanium is used in the aero-space industry and in chemical and marine engineering. Titanium carbide, TiC, is harder even than carborundum (silicon carbide) and finds many industrial uses.

Answer the following questions using the information about titanium contained in the above passage and on pages 198–201.

(a) Suggest a likely type of structure and bonding for the very hard substances, silicon carbide and titanium carbide.

(b) From evidence in the passage what type of bonding do you think exists in titanium chloride? Give your reasons.

From information on pp. 198–201 what formula and type of bonding might have been expected? Give your reasons.

(c) Give an explanation for the fuming in moist air of titanium chloride.

(d) What will be the products of the reaction between $TiCl_4$ and magnesium?

What does this reaction suggest about the position of titanium, relative to magnesium, in the activity series?

Why is the mixture of products from the reaction 'washed with very dilute acid' to leave titanium metal?

(e) Titanium reacts with alkali (as well as slightly with acids). Name another metal which does this. What term is used to describe this property?

(f) What physical property of titanium makes it suitable for use in the aero-space field? Quote your evidence for your answer.

(S.C.E.E.B.)

8 The Alkali Metals

Table 8.1

element	atomic number	atomic mass	electron arrangement
lithium	3	6.9	2,1
sodium	11	23.0	2,8,1
potassium	19	39.1	2,8,8,1
rubidium	37	85.5	2,8,18,8,1
caesium	55	132.9	2,8,18,18,8,1
francium	87	(223)	2,8,18,32,18,8,1

(   ) indicates the mass no. of the most stable isotope.

Occurrence

The alkali metals are very reactive and do not occur naturally in the free state. They frequently occur as halides containing the alkali metal as a cation with a single positive charge.

Lithium. Occurs as a small proportion of certain rare silicates.

Sodium. Large quantities of sodium ions occur in sea water, and the evaporation of former seas has produced beds of 'rock salt'. Deposits of sodium carbonate, sodium borate and sodium nitrate also occur. British sources are Cheshire, Lancashire and Teeside (all as NaCl). The ash of sea weed contains a considerable amount of sodium carbonate ('soda').

Potassium. Potassium ions are present in sea water, and in deposits of the chloride, sylvite, in beds formed by evaporation of earlier seas. The most famous deposits occur at Stassfurt. Recently developed British deposits are in North Yorkshire. Land-plant ash contains potassium carbonate ('potash').

Rubidium and caesium. These are very rare, found as a small proportion of very rare minerals.

Francium. All isotopes are radioactive with short half-lives. It occurs in small quantities in natural radioactive decay series and has been made by nuclear bombardment.

Relative abundance Na > K ≫ Li ≫ Rb > Cs
Sodium and potassium are the fifth and sixth most abundant elements in the earth's crust.

Physical Properties

The alkali metals are all soft and of low density, lithium, sodium and potassium all being less dense than water. Figs. 8.1 to 8.4 illustrate some of the important properties.

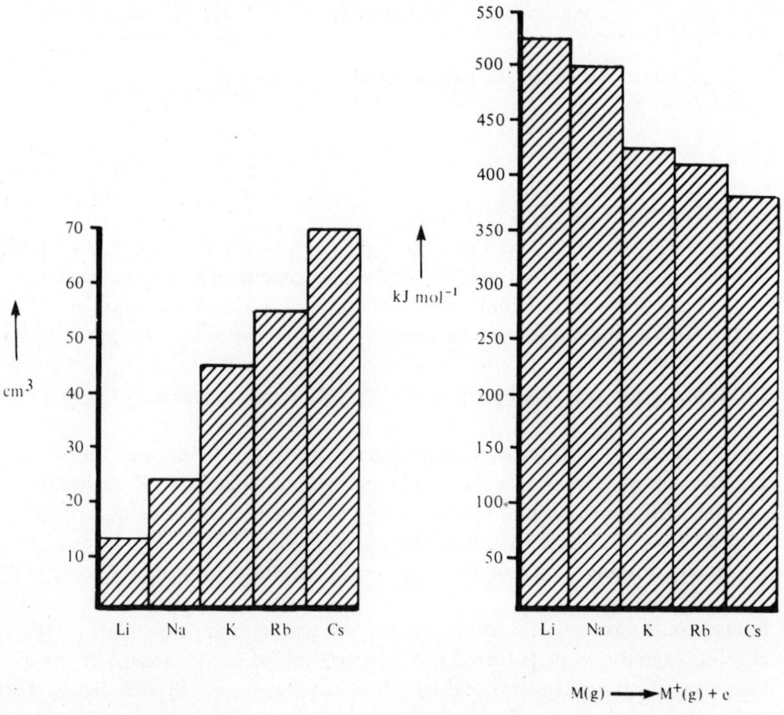

Fig. 8.1 atomic volume

Fig. 8.2 1st ionisation energy

Atomic volume

Predictably, as the atom gains extra layers of electrons, the atomic volume increases.

1st Ionisation energy

As the size of the atom increases, the energy required to remove the outermost electron to form a single positive ion decreases. This energy is the 1st Ionisation Energy for the change

$$M_{(g)} \longrightarrow M^{+}_{(g)} + e$$

At a greater distance from the nucleus, the attraction by the nucleus is obviously less.

The size of the ion produced still depends on the number of electron layers, and hence the **ionic radius** increases down the group.

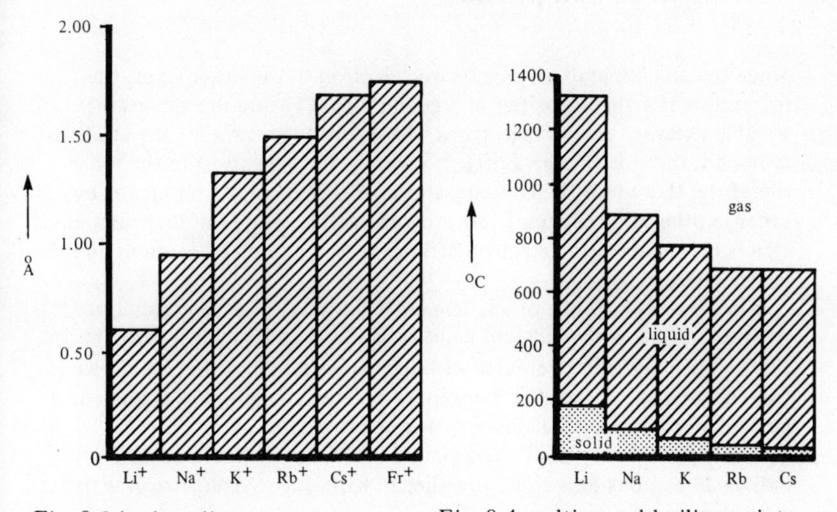

Fig. 8.3 ionic radius

Fig. 8.4 melting and boiling points

Melting point and boiling point

The m.pt. and b.pt. depend on the strength of the forces holding the particles of the element together. These forces in the alkali metal group are those of 'metallic bonding'. The atoms of the metals lose their outer electrons into a common 'pool' which acts as a negative binding agent between the positive ions remaining. The strength of that binding force will depend on the strength of the attraction of the nuclei for the electrons. As the atomic size increases, the attraction decreases, and the m.pt. and b.pt. decrease also.

Extraction of the alkali metals

As the table of E° values shows, the attempted production of alkali metals by electrolysis of aqueous solutions will yield only hydrogen (except with a mercury cathode), however electrolysis of a molten halide will produce the alkali metal at the cathode.

$$M^+_{(l)} + e \longrightarrow M_{(l)}$$

This is the basis of the modern Downs' process for the extraction of sodium from molten sodium chloride, and of the original preparation of sodium (Davy) by the electrolysis of fused sodium hydroxide.

Formation of Compounds

Since the alkali metals all possess one electron in the outer layer, the formation of a single positive charge ion would ensure the achievement of a stable external octet of electrons. However the removal of the electron is an endothermic change as Fig. 8.5 shows. In order to form the ion therefore, there must be compensation for this energy requirement by other exothermic changes. In forming the solid compound this compensation is by means of the **electron affinity** of the non-metal element and the **lattice energy.**

The electron affinity of an element is the amount of energy released when the atom of the element gains an electron in forming a negative ion. The electron affinity decreases with increasing size of atom, (as shown in Fig. 9.5 on page 95 for the halogens) since the attraction of the positive nucleus decreases with distance of the electron from it.

The lattice energy of the compound is the energy liberated when the individual gaseous ions come together to form the crystal lattice of the solid compound.

The energy changes involved in the production of an alkali metal halide are illustrated for sodium chloride, in the diagram of the 'Born-Haber Cycle', Fig. 8.5.

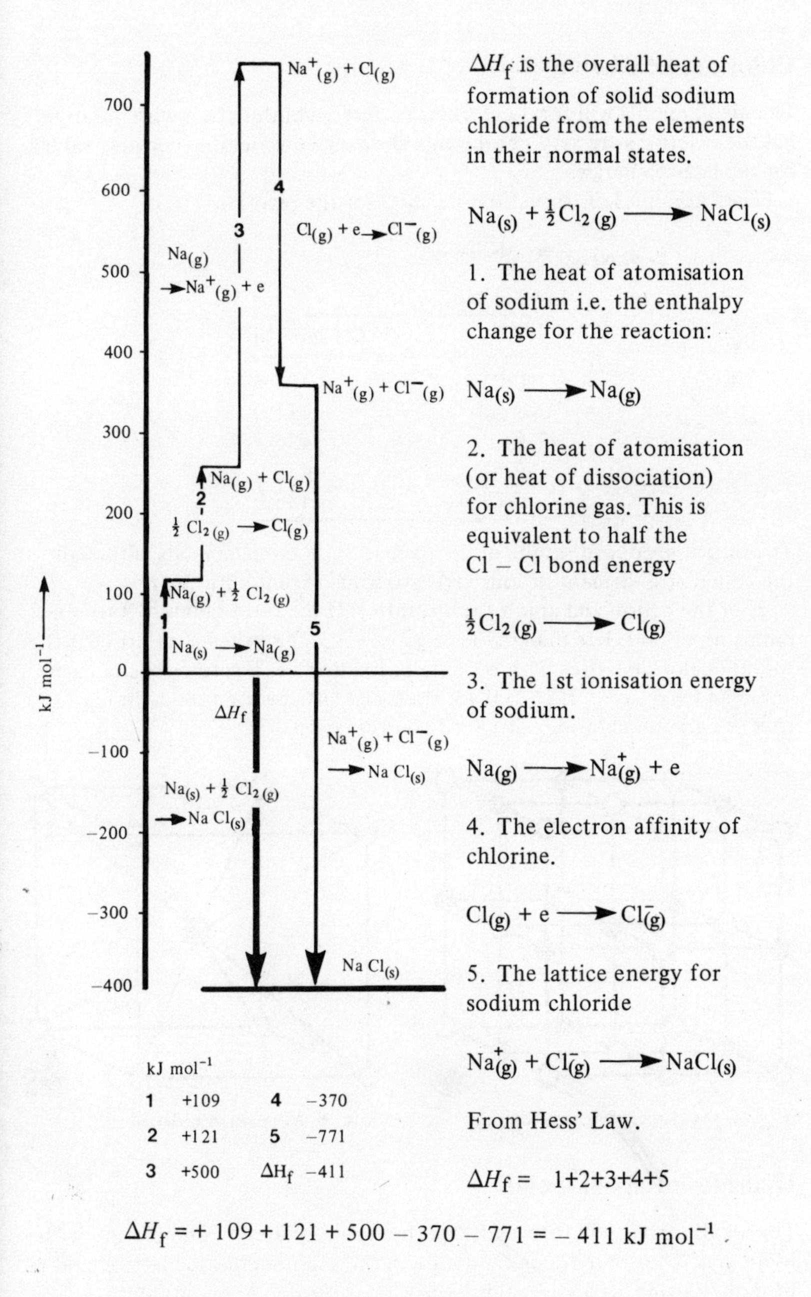

ΔH_f is the overall heat of formation of solid sodium chloride from the elements in their normal states.

$$Na_{(s)} + \tfrac{1}{2}Cl_{2\,(g)} \longrightarrow NaCl_{(s)}$$

1. The heat of atomisation of sodium i.e. the enthalpy change for the reaction:

$$Na_{(s)} \longrightarrow Na_{(g)}$$

2. The heat of atomisation (or heat of dissociation) for chlorine gas. This is equivalent to half the Cl – Cl bond energy

$$\tfrac{1}{2}Cl_{2\,(g)} \longrightarrow Cl_{(g)}$$

3. The 1st ionisation energy of sodium.

$$Na_{(g)} \longrightarrow Na^+_{(g)} + e$$

4. The electron affinity of chlorine.

$$Cl_{(g)} + e \longrightarrow Cl^-_{(g)}$$

5. The lattice energy for sodium chloride

$$Na^+_{(g)} + Cl^-_{(g)} \longrightarrow NaCl_{(s)}$$

From Hess' Law.

$$\Delta H_f = 1+2+3+4+5$$

$$\Delta H_f = +\,109 + 121 + 500 - 370 - 771 = -\,411 \text{ kJ mol}^{-1}.$$

Fig. 8.5 energy changes in the formation of solid sodium chloride from sodium and gaseous chlorine.

Chlorides of alkali metals

The alkali metals will form chlorides by direct combination with chlorine. All the reactions are very exothermic showing considerable negative values for the lattice energies.

The enthalpies of formation, i.e. ΔH_f for the reaction:

$M_{(s)} + \frac{1}{2}Cl_{2\,(g)} \longrightarrow MCl_{(s)}$ are

Table 8.2

	(kJ mol^{-1})
LiCl	– 408.8
NaCl	– 411.0
KCl	– 435.9
RbCl	– 430.6
CsCl	– 433.0

The lattice produced results in the formation of cubic crystals, although the actual arrangement of ions varies to some extent with the relative sizes of the cation and anion **(radius ratio)**. If the ratio **radius of cation/ radius of anion** is less than 0.73, i.e. $\frac{r+}{r-} < 0.73$, then the usual lattice is the sodium chloride lattice with 6 : 6 co-ordination i.e. 6 cations round each anion and *vice versa*. If $\frac{r+}{r-} > 0.73$, the usual lattice structure is similar to that of caesium chloride with 8 : 8 co-ordination.

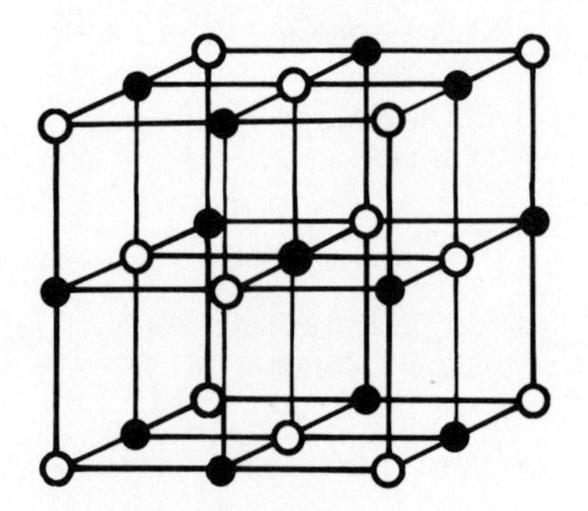

Fig. 8.6 sodium chloride lattice

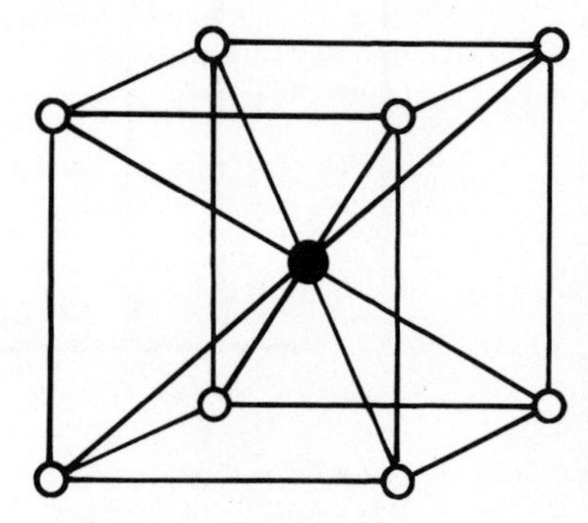

Fig. 8.7 caesium chloride lattice

Hydrides of alkali metals .

The alkali metals will form 'salt-like' hydrides by direct combination with hydrogen. The compounds crystallise in the sodium chloride lattice, and lithium hydride, at least, yields hydrogen at *the anode* on molten electrolysis.

The hydrides contain the H^- ion, and although its formation is endothermic, its formation is compensated for by the lattice energy, as with the halides.

Compare the following:

$$\tfrac{1}{2}H_{2(g)} \longrightarrow H_{(g)} \qquad \Delta H = +218 \text{ kJ mol}^{-1}$$

$$H_{(g)} + e \longrightarrow H^-_{(g)} \qquad \Delta H = -72 \text{ kJ mol}^{-1}$$

i.e. $$\tfrac{1}{2}H_{2(g)} + e \longrightarrow H^-_{(g)} \qquad \Delta H = +146 \text{ kJ mol}^{-1}$$

and $$\tfrac{1}{2}H_{2(g)} \longrightarrow H_{(g)} \qquad \Delta H = +218 \text{ kJ mol}^{-1}$$

$$H_{(g)} \longrightarrow H^+_{(g)} + e \qquad \Delta H = +1316 \text{ kJ mol}^{-1}$$

i.e. $$\tfrac{1}{2}H_{2(g)} \longrightarrow H^+_{(g)} + e \qquad \Delta H = +1534 \text{ kJ mol}^{-1}$$

In fact the formation of H^- ions is less endothermic than the formation of H^+ ions which exist only in solution where they are stabilised by hydration, with the evolution of considerable energy.

The hydrides of the alkali metals all react violently with water, evolving hydrogen and forming sodium hydroxide solution.

$$NaH_{(s)} + H_2O_{(l)} \longrightarrow Na^+_{(aq)} + OH^-_{(aq)} + H_{2(g)}$$

Oxides and hydroxides of the alkali metals

The oxides of the alkali metals are also made by direct combination with oxygen, and are ionic solids with high lattice energy values, containing the O^{2-} ion. With water, all react exothermically giving the corresponding hydroxide.

e.g. $$Na_2O_{(s)} + H_2O_{(l)} \longrightarrow 2Na^+_{(aq)} + 2OH^-_{(aq)}$$

The hydroxides are all strong alkalis, and in the solid state have the sodium chloride lattice structure.

Uses of Alkali Metals

The alkali metals have been chemical curiosities until relatively recently, although they have been used practically to extract other metals e.g. aluminium from bauxite. Currently molten alkali metals are being used as nuclear reactor coolants, and sodium has been investigated (because of its high conductivity and lightness) with a view to its use in high tension power cables. Alkali metal compounds, especially hydroxides and carbonates, find wide application, in the glass and soap industries in particular. The halogen compounds are referred to in the next chapter.

9 The Halogens

Table 9.1

element	atomic number	atomic mass	electron arrangement
fluorine	9	19.0	2,7
chlorine	17	35.5	2,8,7
bromine	35	80.0	2,8,18,7
iodine	53	126.9	2,8,18,18,7
astatine	85	(210)	2,8,18,32,18,7

(   ) This is the mass number of the isotope with the longest known half life.

Occurrence

Halogens are reactive elements and do not occur naturally in the free state. They usually occur as halides (meaning 'salts') in which the halide ions have a single negative charge, thus completing the external electron octet.

Fluorine. This occurs mainly as calcium fluoride, known as fluorite or fluorspar. Another mineral, cryolite, Na_3AlF_6, used in the extraction of aluminium, is much rarer.

Chlorine. Found mainly as chlorides of sodium, potassium and magnesium in 'evaporite' deposits, e.g. Stassfurt (W. Germany), Co. Durham, N. Yorkshire and Cheshire, the chloride ion is the predominant anion in sea water.

Bromine. Found in small quantities in sea water as bromide ion although higher concentrations occur in enclosed seas, such as the Dead Sea.

Iodine. Iodine is found in even smaller quantities in sea water as iodide ion. Certain types of seaweed, especially *laminaria*, concentrate the iodide, which can be extracted by burning to produce ash, known as kelp, and then solution in water. The main source of iodine, however, is as iodate ion, IO_3^-, which is present in nitrate deposits in Chile.

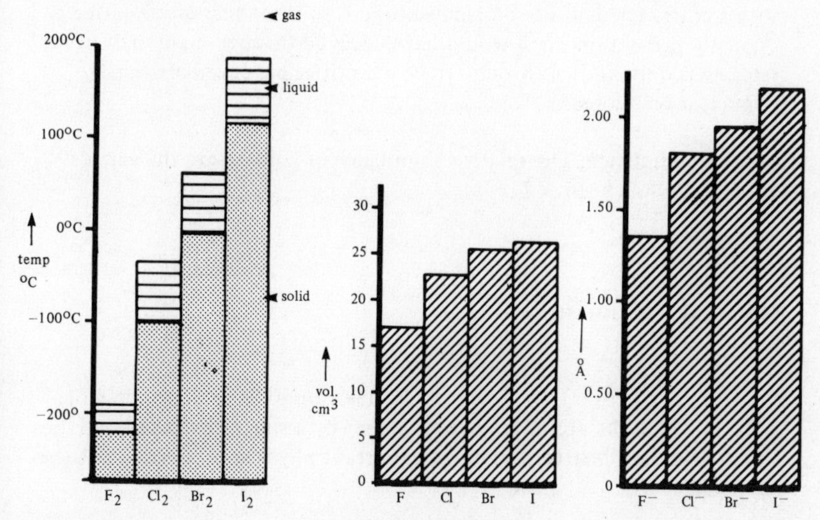

Fig. 9.1 m.pt & b.pt. of halogens

Fig. 9.2 atomic volume (atomic mass/density)

Fig. 9.3 ionic radius

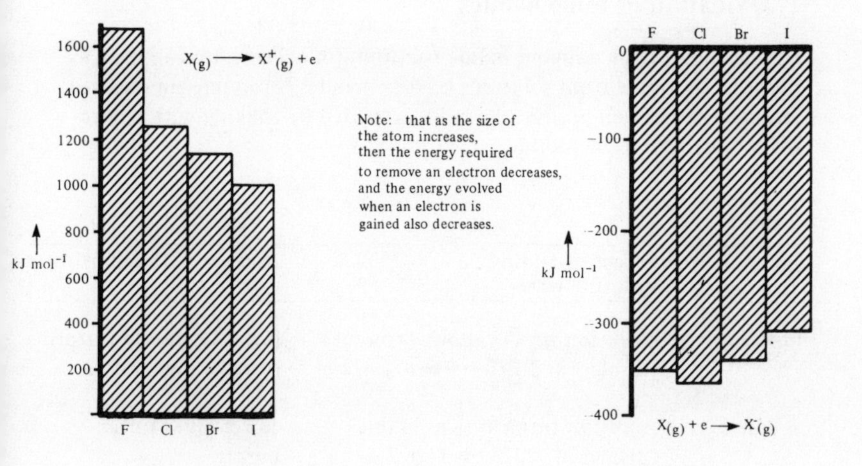

Fig. 9.4 1st ionisation energy

Fig. 9.5 electron affinity

Astatine. This element has not been found in detectable amounts in the earth's crust. The half life of each isotope is so short that any astatine originally present on earth would have decayed through many half lives. Astatine is only known through trace quantities produced artificially by nuclear reactions.

Relative abundance. The relative abundance of halogens in the earth's crust: $F > Cl \gg Br > I$

Physical Properties

When obtained from their compounds, halogens exist as diatomic molecules, in which the atoms are held together by a single covalent bond. Figs. 9.1 to 9.5 illustrate the more important physical properties of these elements.

Obtaining Halogens from their Compounds

1. Oxidation of ionic halides

Displacement from aqueous halide solution. Chlorine water can displace bromine or iodine from solutions of their halides. Bromine can displace iodine. The halogen produced can be extracted by shaking with chloroform or carbon tetrachloride.

Table 9.2

halide ion in solution	effect of adding chlorine water	shaking with chloroform
bromide	solution turns yellow, bromine displaced. $2\ Br^- \longrightarrow Br_2 + 2\ e$	lower layer (chloroform) turns orange-brown
iodide	solution turns brown, iodine displaced. $2\ I^- \longrightarrow I_2 + 2\ e$	lower layer turns purple

In each of these reactions, chlorine is the electron acceptor, $Cl_2 + 2\ e \longrightarrow 2\ Cl^-$. Combining the half equations for the chlorine/bromide reaction, gives the following overall equation. Metal ions present in the solution are spectator ions and hence are omitted.

Loss of electrons – Oxidation (LEO)

$$Cl_2 + 2Br^- \longrightarrow Br_2 + 2Cl^-$$

Reduction by electron gain (REG)

Similar equations can be written for the chlorine/iodide and bromine/iodide reactions.

Using concentrated sulphuric acid and manganese(IV) oxide:

Table 9.3

halide tested	effect of adding conc. H_2SO_4 to halide crystals
iodide	oxidation to iodine $2I^- \longrightarrow I_2 + 2e$
bromide	,, ,, bromine $2Br^- \longrightarrow Br_2 + 2e$
chloride	displaces acid with lower b.pt. $HCl_{(g)}$
fluoride	,, ,, ,, ,, ,, $HF_{(g)}$

Sulphuric acid is the electron acceptor and is reduced to sulphur dioxide and hydrogen sulphide. The acid is not a sufficiently powerful electron acceptor to oxidise chlorides or fluorides, however. The oxidising power can be increased by mixing manganese(IV) oxide, MnO_2, with the halide before adding the sulphuric acid. Under these conditions, a chloride is oxidised to chlorine ($2Cl^- \longrightarrow Cl_2 + 2e$), but fluorine is not obtained from a fluoride.

Incidentally, another manganese compound is used in a common laboratory method for preparing chlorine. Solid potassium permanganate is used to oxidise concentrated hydrochloric acid.

Electrolysis of halide solutions. If an aqueous solution of a sodium or potassium halide is electrolysed with graphite or platinum electrodes, the halogen is produced at the anode and hydrogen gas at the cathode.

At the anode: $2X^- \xrightarrow{\text{LEO}} X_2 + 2e$, where X = Cl, Br or I.

At the cathode: $2H^+ + 2e \xrightarrow{\text{REG}} H_2$

Aqueous fluorides are an exception in that they yield oxygen at the anode.

$$2H_2O \xrightarrow{\text{LEO}} O_2 + 4H^+ + 4e$$

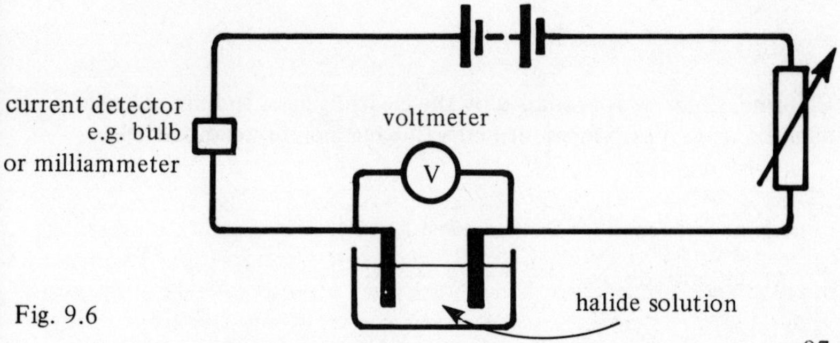

Fig. 9.6

If a voltmeter is connected in parallel with the electrolysis cell as shown in Fig. 9.6, it is possible to measure the minimum voltage that is required to bring about electrolysis. This gives an indication of the relative difficulty experienced in the removal of the extra electron from the halide ion by the anode.

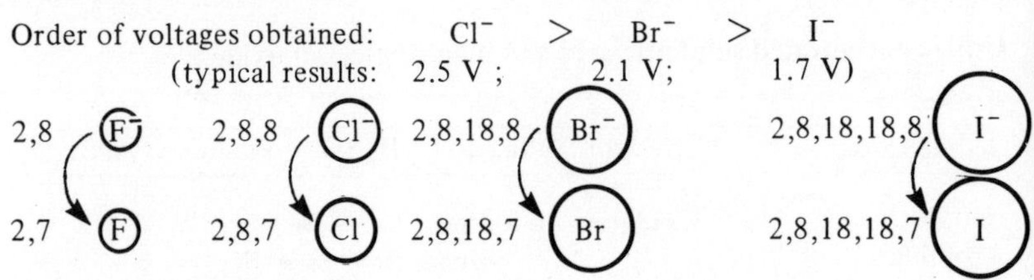

Fig. 9.7

Generally, the extra electron of the halide ion is held least firmly by the nucleus of the largest ion and most firmly in the smallest ion. Hence, iodine is the easiest halogen to obtain, fluorine the most difficult.

The results of the above methods of oxidising halide ions and, in particular, the failure to obtain fluorine, are consistent with predictions from the table of E° values (see page 203). Fluorine is the most powerful oxidising agent in the series, so that chemical oxidation of fluoride to fluorine is not possible. However, chemical oxidation of the other halides is possible as E° values of half-reactions involving the permanganate ion and sulphate ion show.

The E° values also indicate why fluorine cannot be produced by electrolysis of an *aqueous* fluoride. The production of chlorine instead of oxygen from an aqueous chloride is explained in Chapter 5.

2. Special methods

Fluorine. This is produced by the electrolysis of a molten fluoride. A mixture of anhydrous HF and KF is often used. The reaction at the anode is:

$$2F^-_{(l)} \longrightarrow F_{2\,(g)} + 2e$$

Chlorine. Chlorine is produced by the electrolysis of sodium chloride solution using a carbon anode and a flowing mercury cathode. (See Chapter 5 page 51).

At the anode: $2Cl^- \longrightarrow Cl_2 + 2e$

At the cathode: $2Na^+ + 2e \longrightarrow 2Na$

This occurs only with a mercury cathode, otherwise the product is hydrogen. The sodium dissolves in the mercury forming an amalgam which is run off into water in a separate vessel so that the following reaction occurs:

$$2Na + 2H_2O \longrightarrow 2Na^+ + 2OH^- + H_2$$

Hence the process yields hydrogen and sodium hydroxide from common salt solution in addition to chlorine.

Iodine. Sulphite ions are used to bring about the reduction of iodate ions (IO_3^-). If starch is present, a definite time interval elapses before a blue colour appears. The length of time depends on the temperature and concentration of the reacting solutions. This is an example of a 'clock reaction'.

Chemical Reactions of the Halogens

1. Reaction with hydrogen

Fluorine reacts instantly and explosively to form hydrogen fluoride.

Equal volumes of chlorine and hydrogen explode when exposed to sunlight, photoflash or light from burning magnesium. A 'chain reaction' occurs which begins with some of the chlorine molecules being split into separate atoms by light energy.

$$Cl_2 \xrightarrow{\text{light}} Cl^\bullet + Cl^\bullet$$

These atoms or 'free radicals' are very reactive since they each possess an unpaired electron (represented by $^\bullet$). The following reactions occur with great rapidity.

$$Cl^\bullet + H_2 \longrightarrow HCl + H^\bullet$$ – also an atom or free radical

$$H^\bullet + Cl_2 \longrightarrow HCl + Cl^\bullet$$ – available to repeat the process

Bromine reacts slowly with hydrogen in sunlight, faster on heating.

Iodine's reaction is slower still and incomplete.

} The hydrogen halide is formed

Chlorine is capable of reacting with hydrocarbons, e.g. a candle will continue to burn in chlorine with a red smoky flame, forming carbon and hydrogen chloride. Other reactions of halogens with hydrocarbons are dealt with in a later chapter.

Thermal decomposition of hydrogen halides. Hydrogen halides are produced by the action of warm concentrated phosphoric acid on sodium or potassium halides, see Fig. 9.8. The hydrogen halide is then heated as it passes through the pyrex side-tube.

Why is conc. H_2SO_4 unsuitable for producing all the hydrogen halides? (See page 97).

Hydrogen iodide decomposes very readily to form hydrogen and iodine.

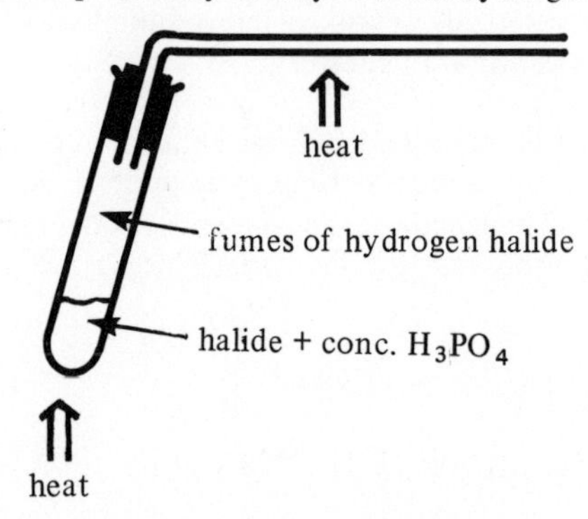

Fig. 9.8

$$2HI \longrightarrow H_2 + I_2$$

Hydrogen bromide decomposes to some extent, HCl negligibly and HF not at all up to at least 1000°C.

These results are in agreement with the hydrogen-halogen bond energies (see page 202). The extent of overlap of electron pair-clouds is relatively greater for a hydrogen atom and the small fluorine atom, than for the hydrogen atom and the much larger iodine atom.

2. Reaction with metals

Dutch metal (an alloy of copper and zinc) in the form of a very thin foil catches fire in contact with chlorine. With bromine and iodine the reaction is much slower and less complete.

Iron, in the form of hot steel wool, burns in chlorine to give iron(III) chloride as a brown smoke. Again the reactions with bromine and iodine vapour are much slower.

Sodium burns in all halogen gases or vapours forming the sodium halides as white smokes. The energy produced is greatest with fluorine and least with iodine.

$$Na \xrightarrow{LEO} Na^+ + e \qquad \tfrac{1}{2}X_2 + e \xrightarrow{REG} X^-$$

(X = a halogen)

Fluorine reacts directly with all metals. It can be stored in copper or stainless steel containers since it forms a thin layer of the fluoride which prevents further corrosion.

3. Reaction with water

Fluorine reacts violently with water, forming hydrofluoric acid and oxygen.

Chlorine forms a pale-green solution having the smell of chlorine. 'Chlorine water', as it is called, gives a positive test for chloride ions, turns blue litmus red before bleaching it and, on standing in daylight, slowly evolves oxygen. Hence chlorine water contains free chlorine molecules, chloride ions, hydrogen ions and something which bleaches and releases oxygen. This is the hypochlorite ion, ClO^-.

$$Cl_2 + H_2O \longrightarrow Cl^- + ClO^- + 2H^+ \quad \left\{ \begin{array}{l} \text{A mixture of hydrochloric} \\ \text{and hypochlorous acids.} \end{array} \right.$$

In bleaching, the ClO^- ions remove electrons from the colouring matter which is thereby oxidised. This contrasts with SO_2, which bleaches by reduction (donating electrons).

$$ClO^- + 2H^+ + 2e \longrightarrow Cl^- + H_2O$$

In daylight, oxygen is evolved and the solution changes to hydrochloric acid.

$$2\,ClO^- \longrightarrow 2Cl^- + O_2$$

Bromine reacts with water in a similar manner to chlorine. The hypobromite ion, BrO^-, also possesses bleaching properties although less markedly than ClO^-.

$$Br_2 + H_2O \longrightarrow Br^- + BrO^- + 2H^+$$

Iodine is scarcely soluble in water and the solution has no bleaching action. It is soluble in an aqueous iodide solution.

4. Reaction with alkalis

Chlorine reacts with cold NaOH solution to form a mixture of sodium chloride and sodium hypochlorite. This solution is more stable than chlorine water and is used as a bleach and disinfectant, e.g. 'Domestos', 'Milton'.

$$Cl_2 + 2Na^+ + 2OH^- \longrightarrow 2Na^+ + Cl^- + ClO^- + H_2O$$

With hot NaOH solution, sodium chlorate, $NaClO_3$, an important weed-killer, is formed.

$$3Cl_2 + 6Na^+ + 6OH^- \longrightarrow 6Na^+ + 5Cl^- + ClO_3^- + 3H_2O$$

Tests for Halide ions

1. Chlorine water. Bromide and iodide ions in solution can be detected by adding chlorine water, followed by either chloroform or carbon tetrachloride, and shaking.

Bromide: orange-brown lower layer due to displacement of bromine.
Iodide: purple lower layer due to displacement of iodine.

2. Silver(I) nitrate. The suspected halide solution is acidified with dilute HNO_3 and silver(I) nitrate solution is added. Chloride, bromide and iodide ions yield precipitates which are then treated with dilute ammonia solution.

Precipitation: $Ag^+_{(aq)} + X^-_{(aq)} \longrightarrow AgX_{(s)}$ $\quad X = Cl, Br, I.$

Table 9.4

chloride	bromide	iodide
white ppt. of AgCl, readily soluble in dilute ammonia.	cream ppt. of AgBr, slightly soluble in dilute ammonia.	yellow ppt. of AgI, insoluble in dilute ammonia.

3. Concentrated sulphuric acid in conjunction with manganese(IV) oxide can be used to identify the presence of a halide ion in a solid. See page 97.

Some Important Uses of Halogens and their Compounds

1. Fluorine compounds

Hydrofluoric acid is very corrosive and is used to etch glass. Aluminium is obtained by molten electrolysis of its oxide (m.pt. about 2000°C). The oxide is mixed with cryolite, Na_3AlF_6, to lower the temperature of the melt to about 900°C. Fluorspar, CaF_2, is used as a flux in high grade steel production. 'Arctons' are fluorine-containing carbon compounds used as refrigerants and aerosol propellents. 'Fluothane' is a very important anaesthetic.

$$\begin{array}{ccc} & Cl & \\ & | & \\ F - & C & - F \\ & | & \\ & Cl & \end{array} \qquad\qquad \begin{array}{cccc} & F & Cl & \\ & | & | & \\ F - & C - & C & - Br \\ & | & | & \\ & F & H & \end{array}$$

an 'arcton' 'fluothane'

PTFE is an important non-stick material and is also a good electrical insulator.

2. Chlorine and its compounds

Chlorine is used as bleach and as a disinfectant for drinking water and swimming pools. Many solvents are chlorine compounds, e.g. dry-cleaning fluids such as trichloroethane. Fire extinguishers: many contain carbon tetrachloride or similar compounds. Pesticides: include chlorinated hydrocarbons, e.g. BHC, DDT. Antiseptics: include chlorinated phenols, e.g. TCP, 'Dettol' (see page 176). Polymers: PVC and neoprene (an artificial rubber) also contain chlorine. Chloroethane is used in the manufacture of lead tetra-ethyl which is added to petrol to eliminate 'knocking' in car engines.

3. Bromine compounds

Dibromoethane is added to petrol to remove the lead as volatile lead bromide. Silver bromide, along with other silver compounds, is used in photography.

$$\begin{array}{cccc} & H & H & \\ & | & | & \\ H - & C - & C & - H \\ & | & | & \\ & Br & Br & \end{array}$$

1,2-dibromoethane

4. Iodine and its compounds

Tincture of iodine is of minor use as an antiseptic. Iodine is an essential part of the diet; its absence causes 'goitre'. Iodine compounds are used as radio-opaque dyes so that organs can be made visible under X-rays.

Examples for practice

1. In 1785, the French chemist Berthollet described an experiment which he claimed proved that the recently prepared substance called 'oxymuriatic acid' – now known to be the element chlorine – was a compound containing oxygen. He had exposed a solution of 'oxymuriatic acid' in water to sunlight and after some time found that oxygen gas had collected above the liquid as the sketch shows:

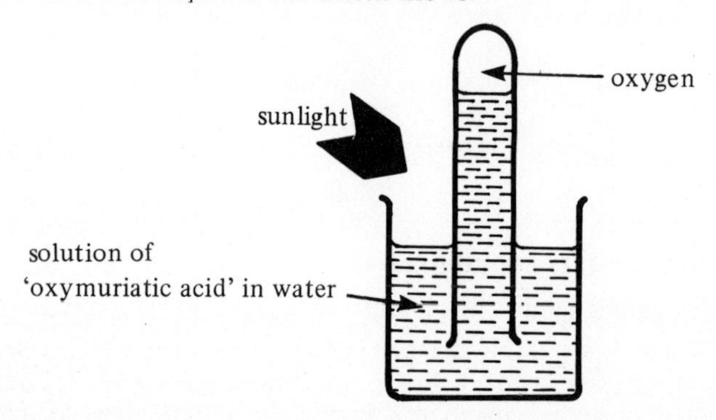

Berthollet found that after the experiment the liquid in the tube was 'muriatic acid', a substance which could also be prepared from common salt and sulphuric acid.

(a) Explain why this experiment does NOT prove that 'oxymuriatic acid' contains oxygen.
(b) What is the modern name for 'muriatic acid'?
(c) Write an equation for the reaction occurring in the tube.
(d) Mention one other reaction involving chlorine which could be initiated by sunlight. (S.C.E.E.B.)

2. You will find first ionisation energies for the elements on pp. 200-201.
(a) What trend in first ionisation energies do you observe within the group of halogen elements?
(b) Representing a halogen atom by X, write an equation to show what is happening during the 'first ionisation' process.
(c) Offer an explanation in terms of atomic structure for the trend observed.
(d) Suggest a reason for the lack of ionisation energy data for the element Astatine (At, atomic number 85). (S.C.E.E.B.)

3. The following table shows the mass numbers of the naturally-occurring isotopes of the halogen elements:

element	mass number of isotopes
fluorine	19
chlorine	35
	37
bromine	79
	81
iodine	127

(a) Use this table and the relevant data on pp. 198-199 to find support for the statement that 'the number of neutrons in the atomic nucleus tends to be even'.
(b) The atomic weight of bromine is 80.0. What information does this give about the isotopes ^{79}Br and ^{81}Br?

(S.C.E.E.B.)

4. (a) The brown liquid iodine monochloride (ICl) is produced when iodine and chlorine are mixed. When a few drops of the brown liquid are added to a gas jar of propene, the brown colour disappears. Name and write an extended structural formula for a compound likely to be produced in this reaction.
(b) In which direction would iodine monochloride be polarised? Would it be more or less polar than iodine monofluoride?

(S.C.E.E.B.)

10 Rate of Reaction

Introduction

In our study of chemical reactions, we have observed a considerable variation in the rate at which those reactions occur. Some reactions are very rapid, e.g. neutralisation of an acid by an alkali and precipitation of an insoluble salt. These reactions involve the interaction of ions which are initially free in solution.

Reactions which involve covalent species are frequently slow, since bonds in the reactant molecules have to be broken before new bonds, and hence new products, can be formed. The fermentation of glucose and the hydrolysis of polysaccharides and proteins are examples of slow reactions. Some molecular reactions become extremely rapid if sufficient energy is supplied to break at least some of the bonds. This is illustrated in the ignition of a hydrogen – oxygen mixture and in the explosion of a hydrogen – chlorine mixture when initiated by light.

In this chapter we shall investigate the factors which influence the rate of a reaction and try to give an explanation of the results.

Changing the Particle Size

1. Phosphorus in air (CARE)

A lump of phosphorus smoulders in air but does not usually catch fire at room temperature. The lump can be broken down into separate molecules by dissolving it in a suitable solvent. When this solution is poured onto a filter paper and the solvent allowed to evaporate, the phosphorus ignites.

2. Acid + carbonate

For example:

$$CaCO_{3(s)} + 2H^+_{(aq)} + 2Cl^-_{(aq)} \longrightarrow Ca^{2+}_{(aq)} + 2Cl^-_{(aq)} + CO_{2(g)} + H_2O_{(l)}$$

Fig. 10.1 illustrates a suitable apparatus for investigating the effect of particle size on the rate of this reaction. The volume and molarity of the acid are kept constant. The total volume of CO_2 released will be the same if the weight of chalk is also kept constant. Fig. 10.2 shows typical results.

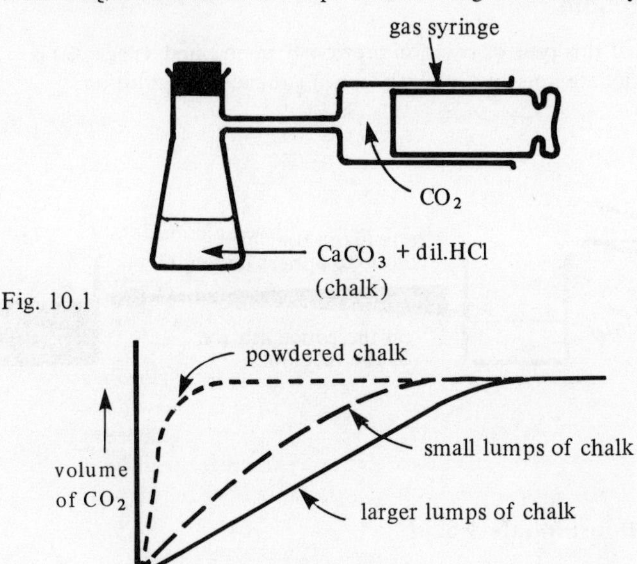

Fig. 10.1

Fig. 10.2

3. Pyrophoric lead

This is a form of very finely divided lead produced by heating lead tartrate in a test tube. When the contents of the tube are poured out into the air, the tiny particles of lead glow brightly as they fall.

Conclusion. These experiments illustrate that the rate of a reaction can be increased by decreasing the particle size of one of the reactants.

Changing the Concentration of a Reactant

Various reactions can be used to study this effect. Some are given below.

1. Magnesium + acid

Measure the time taken for pieces of magnesium ribbon, 1 cm long, to dissolve completely in acid solutions (HCl or H_2SO_4) of varying concentration.

2. Clock reactions

One example of this type of reaction previously mentioned (page 99) is that between iodate ions and sulphite ions in aqueous acid solution.

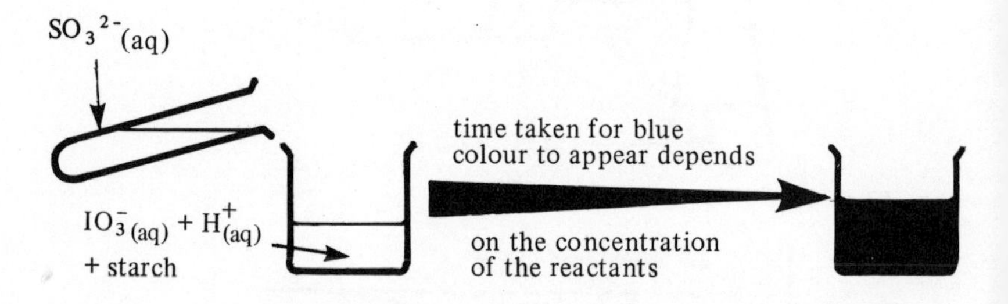

Fig. 10.3

3. Sodium thiosulphate + acid

When acid is added to a solution containing thiosulphate ions, $S_2O_3^{2-}$, a colloidal suspension of sulphur is gradually formed.

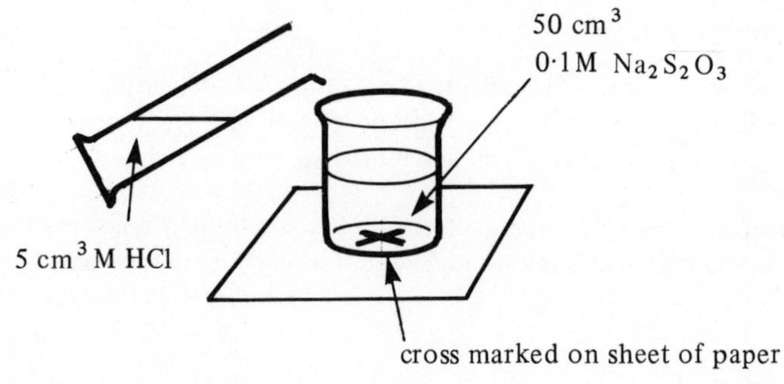

Fig. 10.4

$$S_2O_3{}^{2-}_{(aq)} + 2H^+_{(aq)} \longrightarrow SO_{2\,(g)} + \underset{\text{colloidal}}{S} + H_2O_{(l)}$$

Measure the time, *t*, taken from the point of adding the acid until the cross is obscured by the cloudiness produced. Repeat the experiment by diluting the thiosulphate by a known amount. The total volume of solution and quantity of acid used must be kept constant.

Conclusion: These experiments illustrate that the rate of a reaction can be increased by increasing the concentration of one of the reactants.

In some reactions it is possible to show just how dependent its rate is on the concentration of a particular reactant. For example, in reaction (3) above, a graph of rate against concentration shows a straight line. This indicates that the rate is directly proportional to the concentration of thiosulphate ions.

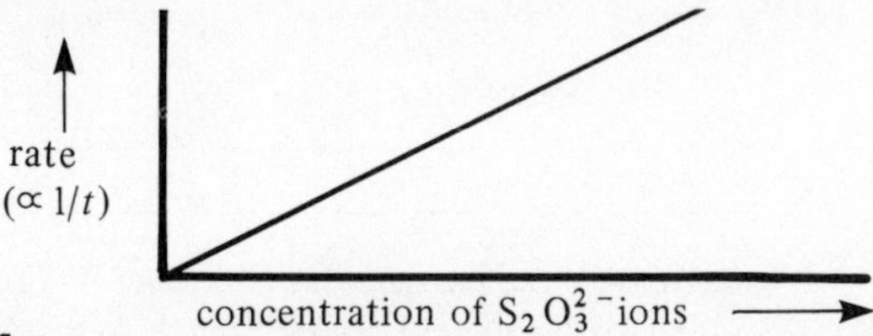

Fig. 10.5

Changing the Temperature

The reactions mentioned in the previous section are also suitable for studying the effect of temperature on the rate of a reaction. This time the concentrations of all of the reactants must be kept constant.

Another suitable reaction for study is that between aqueous solutions of potassium permanganate (acidified with dilute H_2SO_4) and oxalic acid, $(COOH)_2$. The reaction is slow at room temperature but at 80°C decolouration of permanganate ions is virtually instantaneous. The overall reaction is shown in the equation.

$$2MnO_4^-{}_{(aq)} + 5C_2O_4^{2-}{}_{(aq)} + 16H^+_{(aq)} \longrightarrow 2Mn^{2+}_{(aq)} + 10CO_{2\,(g)} + 8H_2O_{(l)}$$

permanganate (purple) — oxalate — manganese(II) (colourless)

Conclusion: The rate of a reaction can be increased by raising the temperature at which the reaction takes place. It is usually found that a small temperature rise can have a considerable effect on the rate of a reaction. Frequently, the rate of a reaction will double for a rise in temperature of around 10°C.

Explaining the Results by means of the Collision Theory

At a very early stage in our study of science, we found that the results of a variety of experiments suggested that matter is made up of very small particles, which we call atoms, ions or molecules. Furthermore, these particles are continually moving, the speed and extent of their motion depending on whether the substance is a gas, a liquid, a solid or in solution. This description is usually referred to as the 'kinetic model of matter'.

For a chemical reaction to occur, the reactants must be brought together in some way so that their particles collide. This is the basic premise of the 'collision theory'. Table 10.1 gives a brief explanation of how the factors, which we have already considered, influence the rate of a reaction by reference to collisions between particles of the reactants.

Table 10.1

factor	explanation
particle size	decreasing the particle size of a solid in a reaction increases the area of contact between reactants. Therefore the number of collisions – and hence rate – increases.
concentration	increasing the concentration of one of the reactants increases the number of collisions between reactant particles.
temperature	raising the temperature makes the particles more energetic and increases the number of collisions per second. (This is an oversimplification as we shall see a little further on.)

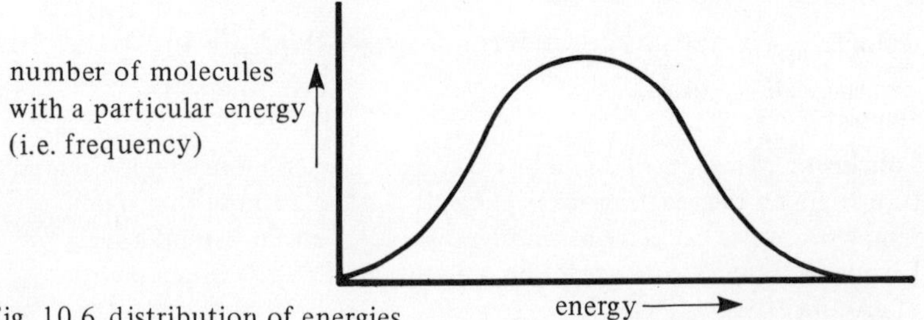

Fig. 10.6 distribution of energies

Reactions occur, then, when reactant particles collide. However, it would appear that not all collisions result in a successful reaction. If they did, all reactions would be instantaneous. Reactions involving gases are often slow at room temperature despite the fact that they mix rapidly by diffusion and that many collisions between molecules will then take place per second.

At a given temperature, the molecules of a gas have widely different energies. Most molecules will have average or nearly average energy, but some will be well below average while others are well above. Fig. 10.6 shows the distribution of energy values. The energy of individual molecules is continually changing as a result of collisions with other molecules. However, the overall distribution of energies remains the same at constant temperature.

Activation energy

In 1889, a Swedish chemist, Arrhenius, suggested that reactions will only occur if the molecules concerned possess a certain minimum energy, which he called the **activation energy**. If in a certain reaction the activation energy is high, only a small proportion of the molecules will possess a high enough energy and hence the reaction will be slow.

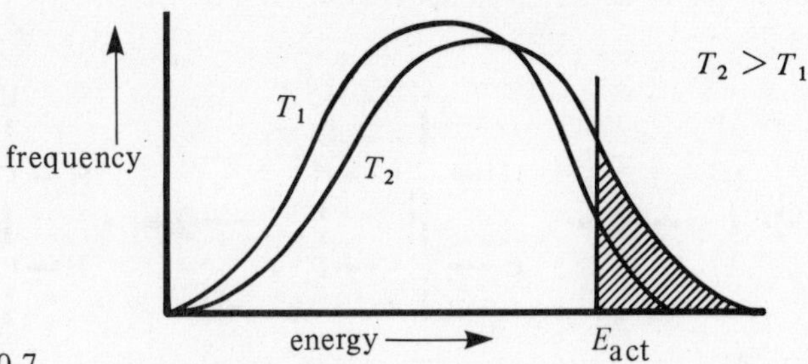

Fig. 10.7

Shaded area represents the no. of molecules with energy equal to or greater than the activation energy, E_{act}.

If the temperature is increased the average energy of the molecules is increased as shown in Fig. 10.7. However, the most significant point is that a small rise in temperature increases considerably the number of those molecules with energy equal to and greater than the activation energy. This is the real reason why a small change in temperature can have such a marked effect on the rate of a reaction.

We can also regard the activation energy as an energy barrier which has to be overcome in going from reactants to products. This is illustrated in the following diagrams.

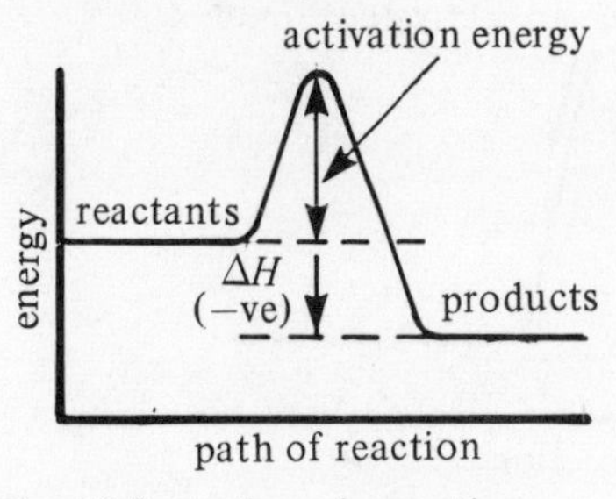

Fig. 10.8 exothermic reaction

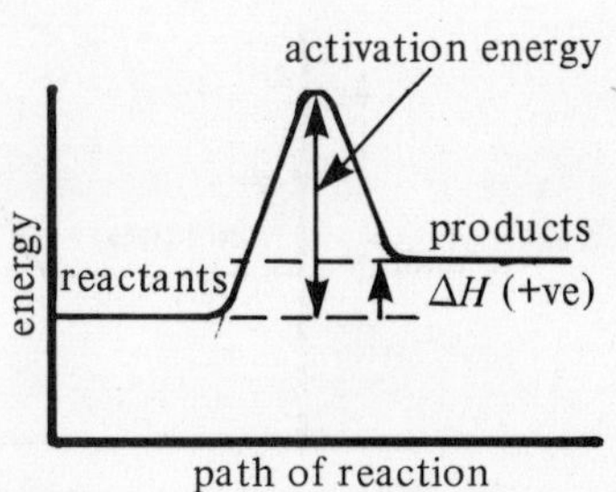

Fig. 10.9 endothermic reaction

The higher the energy barrier, the fewer molecules there are which have sufficient energy to surmount it and, hence, the slower the reaction. It is important to note that the two energy values, i.e. the enthalpy change and the activation energy, are in no way related to each other. The rate of a reaction depends on the activation energy but is independent of the value of ΔH.

Activated complex

In going from reactants to products, an intermediate stage is reached at the peak of the energy barrier at which a highly energetic and unstable particle called an **activated complex** is formed. Such particles exist for only a very short time. The reaction between ethene and hydrogen iodide is believed to go via the activated complex shown in Fig. 10.10.

```
     H                         H                      H
     |                         |                      |
 H — C        H           H —  C ---- H          H —  C — H
     ||   +   |    →           ¦      ¦     →         |
 H — C        I           H —  C ---- I          H —  C — I
     |                         |                      |
     H                         H                      H

 reactants                 activated              product
                           complex (X)            (iodoethane)
```

Fig. 10.10

At the top of the energy barrier, depending on which of the partial bonds are broken, the complex will lose energy and either yield the product or reform the reactants, see Fig. 10.11. This illustrates another very important point about collisions between reactant particles, namely, the angle of collision between molecules may be critical. In this case side-on collision favours formation of the activated complex.

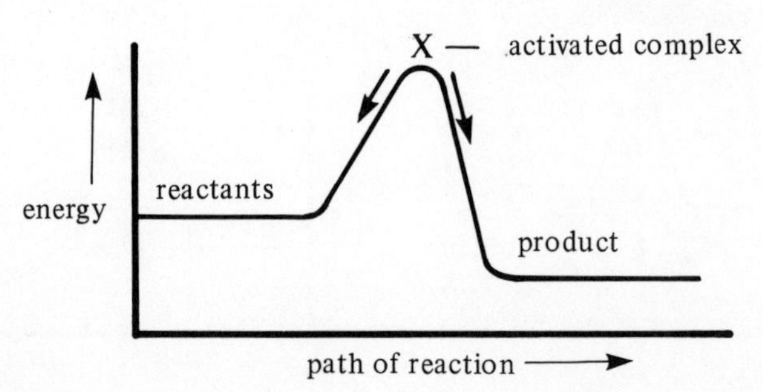

Fig. 10.11

Overcoming the Barrier

Some reactions occur readily at room temperature, e.g. acid-alkali neutralisation, precipitation. In such reactions the activation energy is very low. There are very many reactions which require a supply of energy to make them proceed at a reasonably rapid rate. This energy can be supplied in a number of ways.

1. Heat

Heat is often used as a means of getting a reaction under way, e.g.
(a) warming an acid before adding a basic oxide to make a salt,
(b) burning a metal such as magnesium in air or oxygen, and
(c) displacement reactions involving solids such as aluminium and iron(III) oxide.

If such reactions are sufficiently exothermic, the energy released will sustain the reaction once it has begun. Heat is required as an initial 'surge' to overcome the energy barrier.

2. Shock

The detonation of explosives and the firing of a bullet are examples of reactions which are initiated by shock waves.

3. Light

Several reactions require the help of light energy either to initiate or to sustain them. Some of these reactions are very important.

Photosynthesis:

$$6CO_2 + 6H_2O \xrightarrow[\text{chlorophyll (catalyst)}]{\text{light from sun}} C_6H_{12}O_6 + 6O_2$$

Photography: Silver ions in an unexposed film are reduced in the presence of light to form silver atoms.

$$\underset{\text{colourless}}{Ag^+ + e} \xrightarrow{\text{light}} \underset{\text{black}}{Ag}$$

The darkness of the negative depends on the light intensity and exposure time.

'Blue-printing': A filter paper is soaked in potassium hexacyanoferrate(III) solution and then sprayed with iron(III) citrate solution. An object such as a transparent plastic ruler is placed on the paper and left exposed to sunlight for some time. The marks on the ruler will show up as white against a blue background. The iron(III) ions exposed to the light have been reduced to iron(II) ions which have then reacted with the hexa-cyanoferrate(III) ions to form a substance called Prussian Blue.

$$Fe^{3+} + e \xrightarrow{\text{light}} Fe^{2+}$$

Certain halogen reactions: In the explosion of a hydrogen – chlorine mixture (page 99) and in the substitution reaction between bromine and a hydrocarbon (page 154), the initial steps involve the splitting of halogen molecules into separate atoms. This is brought about by light.

$$X_2 \xrightarrow{\text{light}} X\cdot + \cdot X \qquad (X = \text{Cl, Br})$$

Catalysts

A catalyst is a substance which alters the rate of a reaction without being used up itself. Most catalysts speed up reactions, but there are situations in which catalysts called inhibitors are used to slow down reactions. Inhibitors are used, for example, in rubber as anti-oxidants, in antifreeze to combat corrosion and to stabilise monomers used in the plastics industry.

A simple example of catalysis can be shown using hydrogen peroxide, H_2O_2. A solution of hydrogen peroxide evolves oxygen very slowly even on heating. Oxygen is rapidly evolved when even a small amount of manganese(IV) oxide is added.

$$2H_2O_{2\,(aq)} \longrightarrow 2H_2O_{(l)} + O_{2\,(g)} \quad \text{Catalyst: } MnO_2$$

Catalysts play a very important role in many industrial processes some of which we have discussed in previous years. Table 10.2 summarises some of these processes.

Most of the catalysts dealt with so far are either compounds of metals from the middle of the periodic table or are metals themselves. These metals form a group called the **transition elements** and catalytic behaviour is one of their characteristic properties, along with variable valency and the tendency to form coloured ions.

Furthermore, most of the catalysts listed in table 10.2 are in a different physical state from the reactants, e.g. the use of solid pellets of V_2O_5

Table 10.2

catalyst	process	reaction	importance
vanadium pentoxide	Contact	$2SO_2 + O_2 \rightleftharpoons 2SO_3$	manufacture of sulphuric acid
iron	Haber	$N_2 + 3H_2 \rightleftharpoons 2NH_3$	manufacture of ammonia
platinum	catalytic oxid'n of ammonia	$4NH_3 + 5O_2 \rightleftharpoons 4NO + 6H_2O$	manufacture of nitric acid
nickel	hydrogenation	unsaturated oils + hydrogen $\longrightarrow$ saturated fats	manufacture of margarine
aluminium silicate	catalytic cracking	e.g. $C_{20}H_{42} \longrightarrow 2C_8H_{18} + C + C_3H_6$	manufacture of various fuels and monomers for making plastics

to catalyse the reaction between the gases sulphur dioxide and oxygen. Such catalysts are said to be **heterogeneous.** The surface area of a heterogeneous catalyst is important and it is believed that the reaction occurs on the catalyst surface as shown in Fig. 10.12 (a), (b) and (c)

The Mechanism of Catalysis

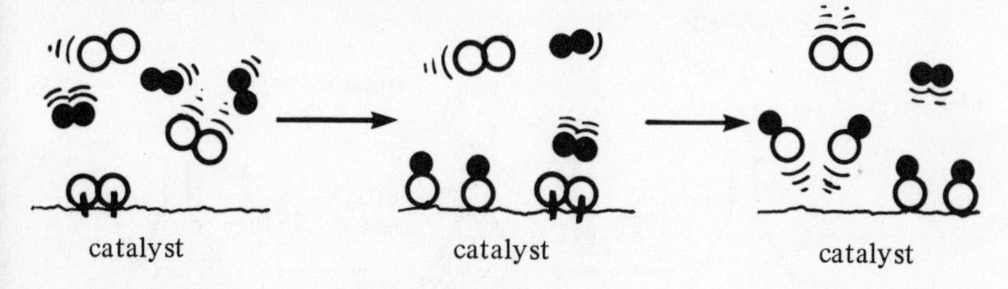

Fig. 10.12

(a) **Adsorption**
Molecules of one or both reactants form bonds with the catalyst. This weakens the bonds *within* the molecules themselves.

(b) **Reaction**
The molecules react on the catalyst surface. The angle of collision is more likely to be favourable since one of the molecules is fixed.

(c) **Desorption**
The product molecules leave the catalyst and the vacant site can be occupied by another reactant molecule.

By weakening the bonds in the molecules of the reactants and by providing a more favourable orientation for the colliding molecules, a catalyst provides a pathway for the reaction which requires much less energy. In other words, a catalyst lowers the activation energy, as shown in Fig. 10.13.

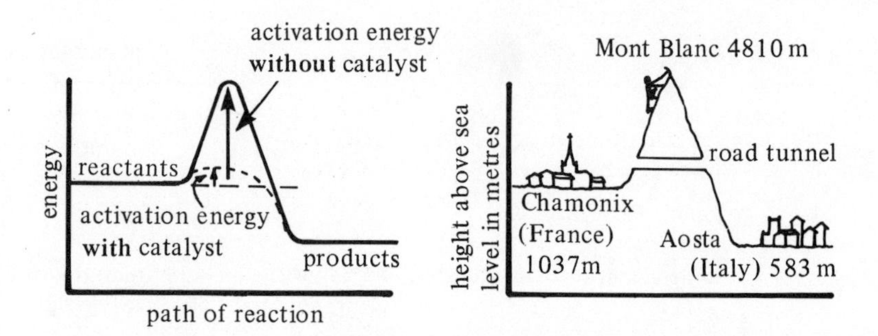

Fig. 10.13 Fig. 10.14

Analogy: (Fig. 10.14) The road tunnel through Mont Blanc provides a less energetic and much quicker route between France and Italy than going 'over the top'. Consequently, the former route is the more frequented!

Although a catalyst is not used up, it may undergo chemical change during its catalytic activity. This is illustrated in the following example.

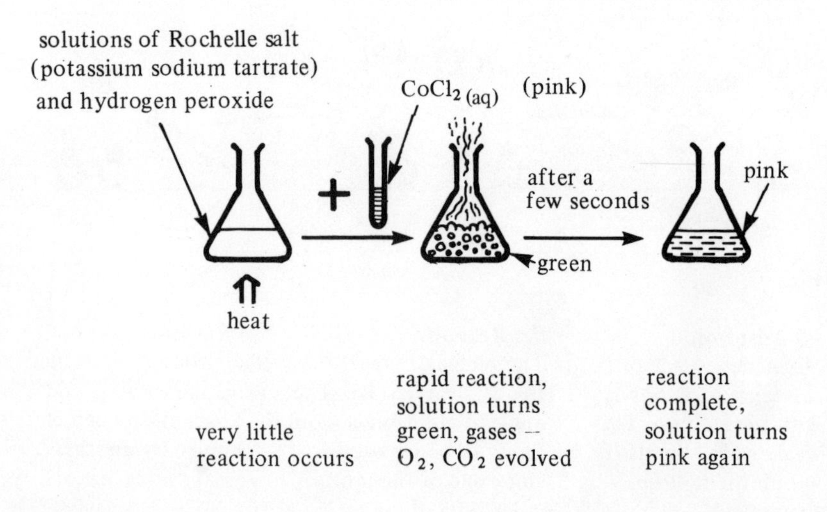

Fig. 10.15

The cobalt(II) ion, Co^{2+}, forms an intermediate complex (possibly involving oxidation to Co^{3+}) during its action as a catalyst but it is re-formed at the end of the reaction. This is also an example of **homogeneous** catalysis since catalyst and reactants are in the same state, i.e. in solution. Other examples of homogeneous catalysts include (i) hydrogen ions in ester formation (page 172), and (ii) haemoglobin, the 'oxygen carrier' present in red blood cells.

In some reactions, one of the products of the reaction may function as a catalyst. This phenomenon is known as **autocatalysis.** This effect can be shown in the reaction between permanganate ions and oxalate ions mentioned earlier in the chapter (page 109). Reduction of permanganate ions, MnO_4^-, to manganese(II) ions, Mn^{2+} is very slow at room temperature but occurs rapidly if a solution containing Mn^{2+} ions is added.

Some catalysts can function for more than one reaction. Platinised asbestos, for example, is frequently used in the laboratory demonstrations of the Contact and Ostwald processes. However, there is no known substance that acts as a universal catalyst. Catalysts which operate for only one reaction are said to be **specific.** Ethanol, C_2H_5OH, vapour is dehydrated to form ethene, C_2H_4, when passed over heated Al_2O_3 catalyst; whereas if a mixture of ethanol vapour and oxygen is passed over a heated Cu catalyst, ethanol is oxidised to form ethanal, CH_3CHO. See pages 156 and 178.

Enzymes are biological catalysts which are highly specific. Molecular shape plays a vital role in their catalytic activity and, as a result, they usually only operate within a fairly narrow range of pH and temperature. Fig. 10.16 illustrates the specific nature of enzymes.

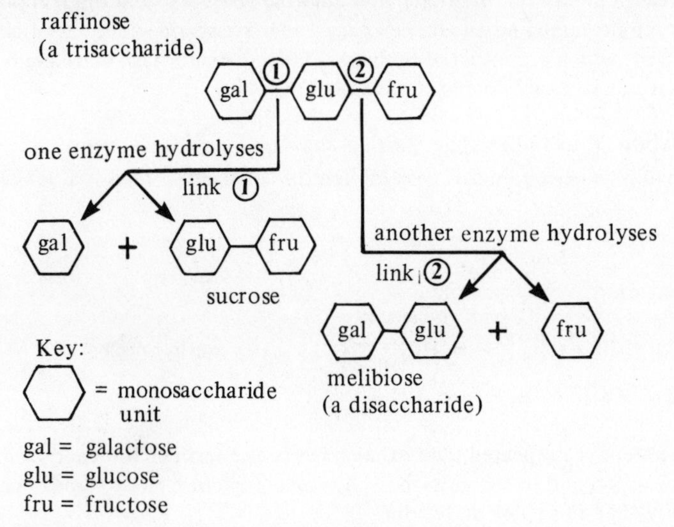

Fig. 10.16

The poisoning of a catalyst can occur in both chemical and biochemical systems. The adsorption of arsenic compounds on the catalyst surface in the Contact process considerably reduces the efficiency of the catalyst. Carbon monoxide attaches itself strongly to haemoglobin thus preventing the uptake of oxygen.

Reaction Mechanism

$$2MnO_4^-{}_{(aq)} + 5\,C_2O_4^{2-}{}_{(aq)} + 16H^+_{(aq)} \longrightarrow 2Mn^{2+}_{(aq)} + 10CO_{2\,(g)} + 8H_2O_{(l)}$$

The above equation states what happens when permanganate ions react with oxalate ions in acid solution. The equation gives information about the chemical nature of the reactants and products. It also tells us something about quantities, i.e. that two moles of MnO_4^- require five moles of $C_2O_4^{2-}$ for complete reaction.

However, the equation does *not* tell us **how** the reaction occurs. Particles must collide to bring about a reaction, but the probability of two permanganate ions, five oxalate ions and 16 hydrogen ions all coming together at the same instant is extremely remote, to say the least. Most reactions are believed to occur by a series of steps in which two, or at the most three, particles are involved. This sequence of intermediate steps is called the **reaction mechanism.**

Examples of reaction mechanisms appear elsewhere in this book, e.g. the reaction between hydrogen and chlorine (page 99) and the reaction between bromine and an alkane (page 154). These are examples of chain reactions which are initiated by light. Chlorine reacts with methane by a similar mechanism which is detailed below.

Initiation: Free radicals (i.e. particles having unpaired electrons) are formed in this step, in this case by light of sufficiently high frequency.

$$Cl_2 \longrightarrow Cl^\bullet + {}^\bullet Cl$$

Propagation:

(1) $Cl^\bullet + CH_4 \longrightarrow HCl + CH_3{}^\bullet$ – a methyl radical

(2) $CH_3{}^\bullet + Cl_2 \longrightarrow CH_3Cl + Cl^\bullet$

These steps are repeated many times, hence the term chain reaction. In this reaction and in the chlorine – hydrogen reaction these steps occur so rapidly that an explosion results.

Termination: The chain reaction can be stopped if the free radicals combine. In this reaction the following terminating steps can occur.

$$CH_3^{\bullet} + CH_3^{\bullet} \longrightarrow C_2H_6$$

$$Cl^{\bullet} + Cl^{\bullet} \longrightarrow Cl_2$$

$$CH_3^{\bullet} + Cl^{\bullet} \longrightarrow CH_3Cl$$

The decomposition of hydrogen peroxide gives us another example of a reaction mechanism. This is thought to occur in two major steps as follows.

$$(1) \quad H_2O_2 \xrightarrow{\text{slow}} H_2O + \underset{\text{atom}}{O}$$

$$(2) \quad \underset{\text{atoms}}{O + O} \xrightarrow{\text{fast}} \underset{\text{molecule}}{O_2}$$

The slowest step in a reaction mechanism is called the **rate – determining step**, since the rate at which it occurs controls the overall rate of reaction. Any effect which will accelerate this step (e.g. the use of MnO_2 as a catalyst in the reaction quoted above) will correspondingly increase the rate of the whole reaction.

Examples for practice

1.

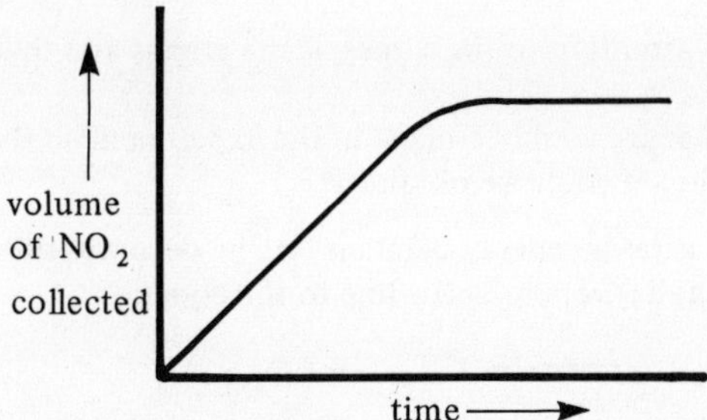

The graph shows how the volume of NO_2 released increases with time when 2g of copper turnings react with excess concentrated nitric acid. Copy the graph (no graph paper required) and add corresponding graphs for the reaction between concentrated nitric acid and

(a) 1 g of copper powder and

(b) a 2 g piece of copper foil.

Label each graph clearly. Pay particular attention to the slope of the graph and the final volume of NO_2.

2.

$$2\,H_2O_2 \longrightarrow 2\,H_2O + O_2$$

Hydrogen peroxide decomposes according to the above equation. Many substances, such as manganese(IV) oxide, lead(IV) oxide and iron(III) oxide function as catalysts in this reaction. Draw a neat diagram of an apparatus you would use to study the effectiveness of these catalysts. Describe briefly how you would use the apparatus to show this.

3. The graph shows the rate at which hydrogen is evolved when excess granulated zinc is added to 100 ml M sulphuric acid at room temperature.

Copy the graph (no graph paper required) and add corresponding graphs for the reactions of excess granulated zinc at room temperature with –

(a) 200 ml M sulphuric acid, and
(b) 100 ml M hydrochloric acid.

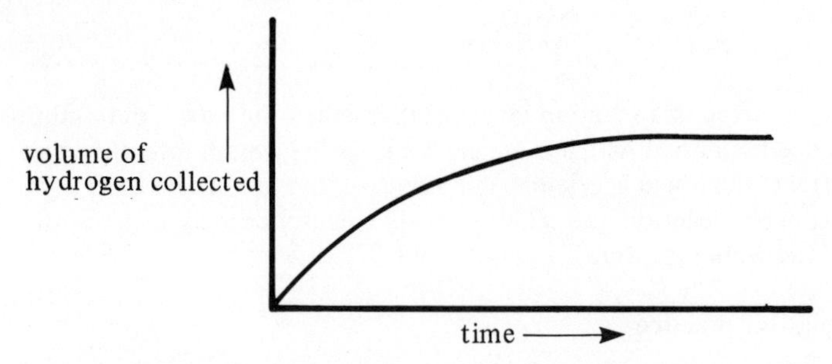

Pay particular attention to the slopes of the graphs and the final volumes of hydrogen.

(c) What other changes could be made in the experiment so that a steeper slope than that shown could be obtained?

4. The graph indicates an energy diagram for the decomposition of ethanal (acetaldehyde) vapour according to the equation:

$$CH_3CHO_{(g)} \longrightarrow CH_{4(g)} + CO_{(g)}$$

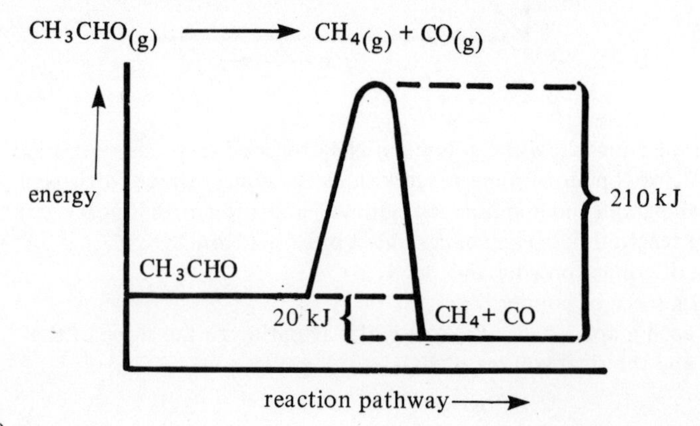

(a) What is the value for the activation energy of the reaction?
(b) What is the enthalpy change for the reaction? Include the correct sign. Is the reaction exothermic or endothermic?
(c) Iodine vapour catalyses the above reaction. Copy the above graph (no graph paper required) and on it indicate by means of a dotted line (.....) the reaction pathway for a catalysed reaction.

11 Chemical Equilibrium

Reversible Reactions

Two examples of reactions which can be readily shown to be reversible are quoted below.

1. Copper(II) sulphate

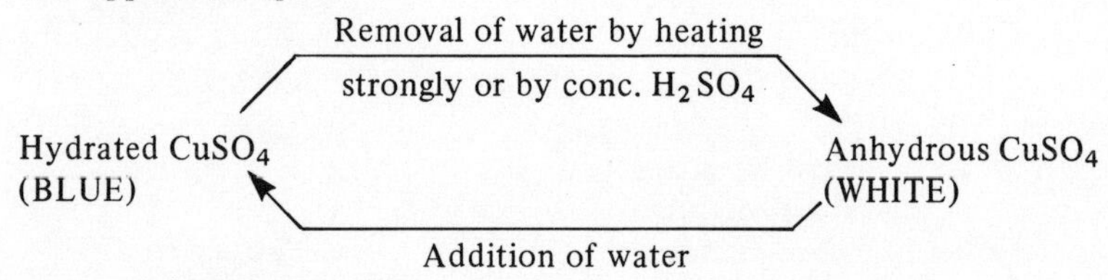

2. Ammonium chloride

$$NH_4Cl_{(s)} \longrightarrow NH_{3(g)} + HCl_{(g)}$$

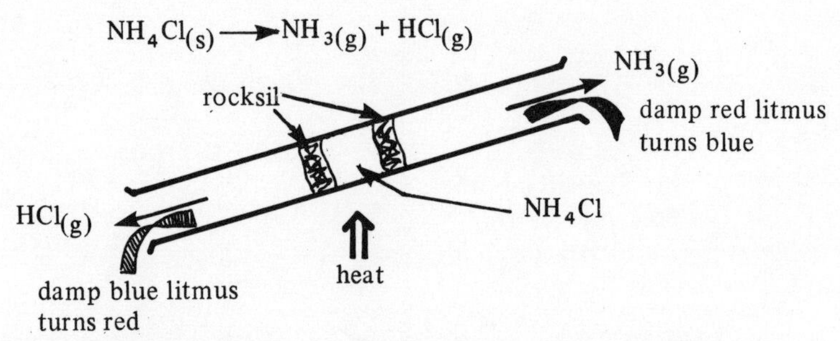

Fig. 11.1 decomposition to yield ammonia and hydrogen chloride.

$$NH_{3(g)} + HCl_{(g)} \longrightarrow NH_4Cl_{(s)}$$

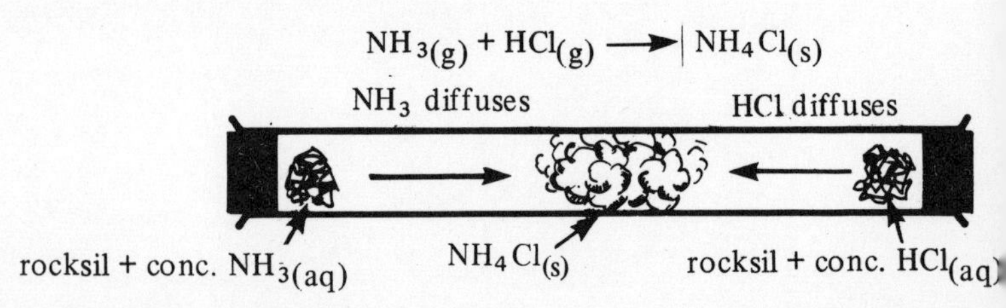

Fig. 11.2 formation by direct combination of ammonia and hydrogen chlori

These are simple examples of reversible reactions. In such reactions, the products of one of the reactions are capable of reacting with each other to reform the original substance(s).

Chemical Equilibrium

Under any fixed set of conditions, a reversible reaction will eventually reach a state of **chemical equilibrium.**

$$A + B \rightleftharpoons C + D$$

If substances A and B are mixed together, the forward reaction will be fast to begin with, but the rate will decrease as the concentrations of A and B decrease. On the other hand, the reverse or backward reaction will have zero rate initially but will increase as the concentrations of C and D increase. Equilibrium is reached when the rate of the forward reaction equals the rate of the reverse reaction. At this point, products are being formed as fast as reactants are being reformed, i.e. there is no overall change in the concentration of reactants and products.

It is important to realise, however, that the reaction does *not* stop when equilibrium is attained. When a saturated solution of a salt such as NaCl is formed, an equilibrium is set up in which as many ions are passing into solution as are being redeposited on the solid crystals, i.e. the rate of solution equals the rate of precipitation. At equilibrium these processes do not cease. Accordingly, chemical equilibrium is described as being **dynamic.**

It is also important to note that when equilibrium is reached this does not imply that the equilibrium mixture consists of 50% reactants and 50% products. This will only very rarely be the case. The actual position of equilibrium can be influenced by a number of factors as we shall see later in this chapter.

Under similar conditions, the same equilibrium can be arrived at from two different starting points. This can be shown (Fig. 11.3) using the

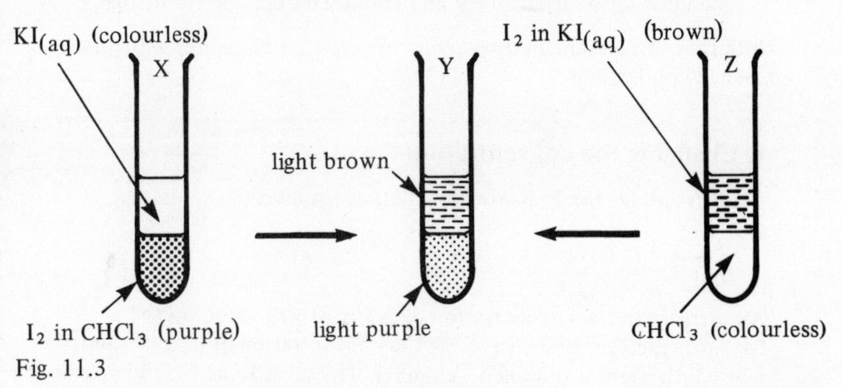

Fig. 11.3

fact that iodine is soluble in trichloromethane (chloroform, $CHCl_3$) and also in aqueous potassium iodide solution. Tubes X and Z represent the two starting positions, in which, in X, iodine is dissolved in $CHCl_3$ only, and in Z it is dissolved in KI solution only. Tube Y represents the equilibrium mixture which is obtained. Equilibrium can be attained quickly by shaking the tubes or slowly by allowing them to stand. Although many reactions appear to be irreversible, e.g. neutralisation of a strong acid and strong base or precipitation such as

$$Ag^{+}_{(aq)} + Cl^{-}_{(aq)} \longrightarrow AgCl_{(s)}$$

all reactions are, in fact, reversible to at least some extent. We can say, however, that in such reactions the equilibrium lies so far in one direction that for practical purposes they can be considered as having gone to completion.

Changing the Position of Equilibrium

We shall now consider the effects which can alter the position of equilibrium in a reversible reaction. Equilibrium is reached when the opposing reactions occur at equal rate. Hence, we should expect that any condition which changes the rate of one reaction more than the other should change the position of equilibrium, i.e. the relative proportion of reactants and products in the mixture.

This section deals with the influence of changing the concentration, the pressure and the temperature on the equilibrium position. The effect of these changes can be summarised by Le Chatelier's Principle which states that

> 'If a system at equilibrium is subjected to any change, the system readjusts itself to try and counteract the applied change.'

Note that this statement *only* refers to reversible reactions which have reached equilibrium.

1. Changing the concentration

Let us consider the following reaction at equilibrium.

$$A + B \rightleftharpoons C + D$$

An increase in the concentration of A (or B) will speed up the forward reaction, thus increasing the concentration of C and D until a new equilibrium is obtained. A similar effect can be achieved by reducing the concentration of Ċ (or Ḋ) in some way. These results agree with Le Chatelier's Principle, since the equilibrium has moved to the right to counteract the applied change. The following reactions can be used to illustrate these points.

Bromine water: Bromine dissolves in water forming a red-brown solution which contains a mixture of Br_2 molecules (responsible for the colour), H_2O molecules, H^+, Br^- and BrO^- ions as shown in the equation.

$$Br_2 + H_2O \rightleftharpoons 2H^+_{(aq)} + Br^-_{(aq)} + BrO^-_{(aq)}$$

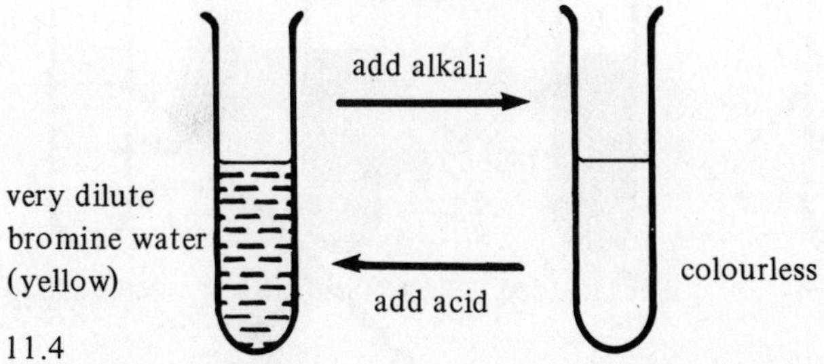

add alkali

very dilute
bromine water
(yellow)

add acid

colourless

Fig. 11.4

The equilibrium position can be adjusted as shown in Fig. 11.4. The addition of OH^- ions removes H^+ ions to form water and the equilibrium shifts to the right. Adding H^+ ions moves the equilibrium back to the left.

Iodine chlorides: When chlorine gas is passed into a tube containing crystals of iodine, the first product is a brown liquid called iodine monochloride, ICl. Excess chlorine converts this substance in a reversible reaction to yellow crystals of iodine trichloride, ICl_3. The equation for this reaction is:

$$\underset{\text{brown liquid}}{ICl_{(l)}} + Cl_{2\,(g)} \rightleftharpoons \underset{\text{yellow crystals}}{ICl_{3\,(s)}}$$

Inverting the tube removes chlorine gas because its density is greater than that of air and the equilibrium adjusts to the left. Thus the yellow crystals disappear and liquid ICl reappears. If chlorine is added, the equilibrium shifts to the right again and yellow crystals re-form. These effects can be repeated many times.

Iron(III) ions + thiocyanate ions (CNS^-): When separate solutions containing iron(III) ions and thiocyanate ions respectively are mixed, a deep blood-red solution is formed due to the presence of complex ions such as $[FeCNS]^{2+}$. This reaction is reversible.

$$Fe^{3+}_{(aq)} + CNS^{-}_{(aq)} \rightleftharpoons [FeCNS]^{2+}_{(aq)}$$

pale yellow colourless red

The intensity of the colour may be taken as an indication of the position of equilibrium. Some of the blood-red solution is diluted until an orange colour is obtained, and this solution is poured into four test tubes till each is half-full. Tube A is kept for reference; crystals of iron(III) chloride, potassium thiocyanate and sodium chloride are added to tubes B, C and D respectively. The results are shown in Fig. 11.5.

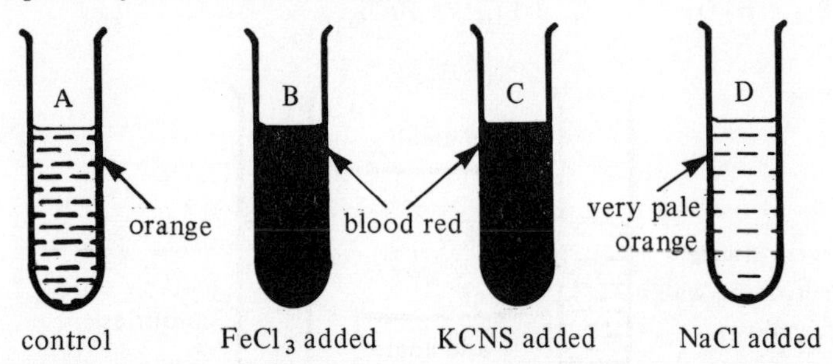

Fig. 11.5

An increase in the concentration of Fe^{3+} ions or CNS^{-} ions shifts the equilibrium to the right and results in the formation of more complex ions. The addition of NaCl removes Fe^{3+} ions due to complex formation with Cl^{-} ions and the equilibrium shifts to the left to compensate.

Weak acids and bases: These substances are only partially dissociated in solution, i.e. undissociated molecules are in equilibrium with free ions.

Weak acid e.g. ethanoic acid

$$CH_3COOH \rightleftharpoons CH_3COO^{-}_{(aq)} + H^{+}_{(aq)}$$

If some solid sodium ethanoate is added, the equilibrium moves to the left since the concentration of CH_3COO^{-} ions has been increased. Thus H^{+} ions are removed and the pH increases.

Weak base e.g. ammonia solution

$$NH_3 + H_2O \rightleftharpoons NH^{+}_{4(aq)} + OH^{-}_{(aq)}$$

If some solid ammonium chloride is added, the equilibrium moves to the left since the concentration of NH_4^{+} ions has been increased. Thus OH^{-} ions are removed and the pH decreases.

2. Changing the temperature

In a reversible reaction, if the forward reaction is exothermic, then the reverse reaction must, of course, be endothermic. If a system at equilibrium is subjected to a change in temperature, the equilibrium position will adjust itself to counteract the applied change, according to Le

Chatelier's Principle. Thus, a rise in temperature will favour the reaction which absorbs heat, i.e. the endothermic process, and a fall in temperature will favour the exothermic reaction. This can be seen in the following examples.

Solubility of ionic compounds: As mentioned earlier in the chapter, in a saturated solution there is an equilibrium between ions trapped in the crystal lattice and hydrated ions in solution, as shown in Fig. 11.6.

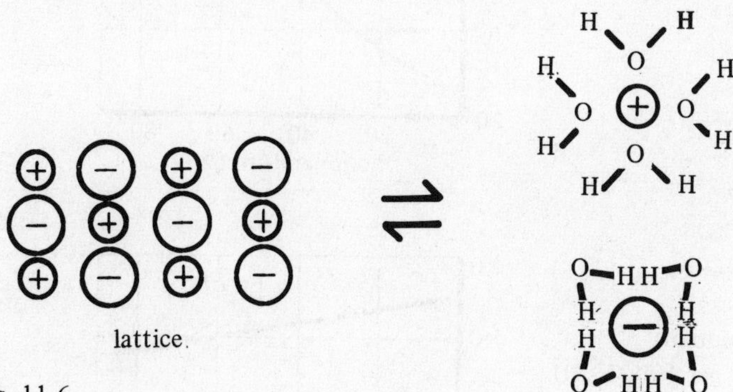

Fig. 11.6

Whether heat is absorbed or released when a substance dissolves in water will depend on the energy required to break the lattice (endothermic), and the energy released when the ions are hydrated (exothermic).

(a) $KNO_{3(s)} \rightleftharpoons K^{+}_{(aq)} + NO^{-}_{3(aq)}$

The forward reaction, i.e. the process of solution, is endothermic ($\Delta H = +35$ kJ mol^{-1}). As the temperature is raised, more salt dissolves, thereby absorbing heat. Hence, the solubility of KNO_3 increases with rising temperature, Fig. 11.7.

(b) $Li_2CO_{3(s)} \rightleftharpoons 2Li^{+}_{(aq)} + CO^{2-}_{3(aq)}$

This time the forward reaction is exothermic ($\Delta H = -17.6$ kJ mol^{-1}). Solubility of this salt decreases with rising temperature, Fig. 11.8.

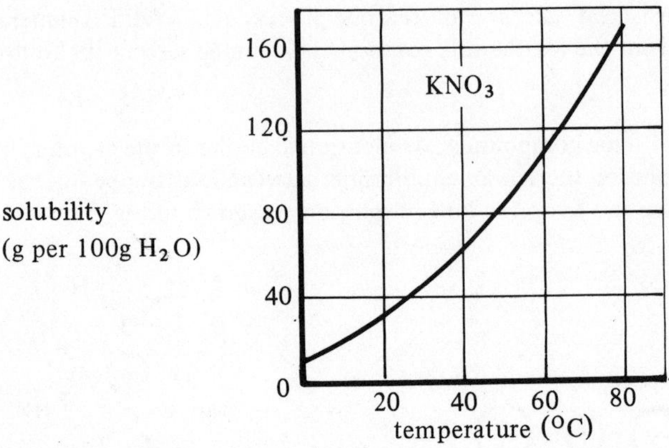

Fig. 11.7

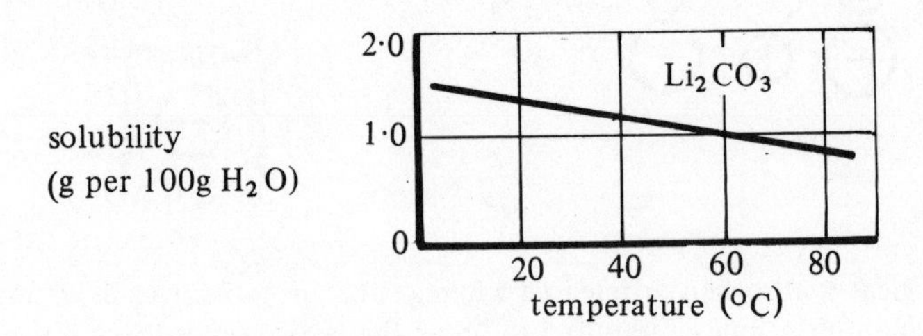

Fig. 11.8

Nitrogen dioxide: Brown fumes of nitrogen dioxide are formed when most metal nitrates are decomposed thermally or when copper is added to concentrated nitric acid. The gas produced is, in fact, an equilibrium mixture of nitrogen dioxide, NO_2 – a dark brown gas, and dinitrogen tetroxide, N_2O_4, which is a colourless gas. This is represented in the following equation. The forward reaction is endothermic.

$$\underset{\text{colourless}}{N_2O_{4(g)}} \rightleftharpoons \underset{\text{dark brown}}{2\,NO_{2(g)}}$$

Fig. 11.9 illustrates the results obtained on subjecting samples of this gas mixture to different temperature conditions. An increase in temperature favours the endothermic reaction and so the proportion of NO_2 increases and the gas mixture becomes darker in colour. A drop in temperature favours the exothermic reaction and, hence, the gas mixture lightens in colour.

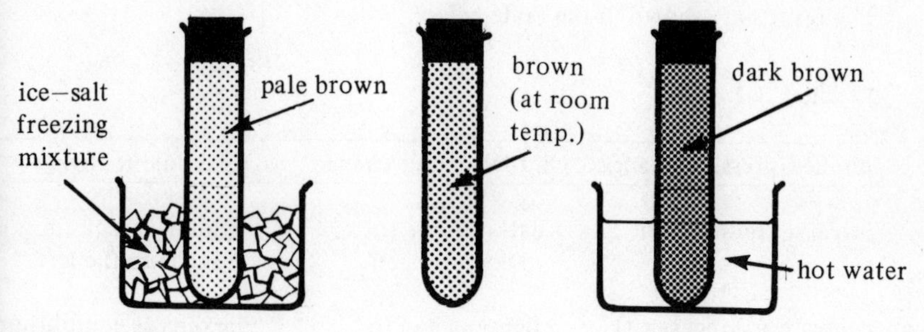

Fig. 11.9

3. Changing the pressure

The pressure exerted by a gas is caused by the freely moving molecules bombarding the walls of the containing vessel. An increase in the number of molecules will be accompanied by an increase in pressure, the size of the container being kept constant. The effect of changes in pressure on a system involving gases is equivalent to the effect of changes in concentration on a system in solution.

The N_2O_4 – NO_2 system is a suitable example to study in this connection.

$N_2O_{4\,(g)}$ $\rightleftharpoons$	$2\ NO_{2\,(g)}$	
1 mole	2 moles	
1 volume	2 volumes	(at same T, P – Avogadro's Law)

An increase in pressure will cause the system to re-adjust to counteract this effect, i.e. it will attempt to reduce the pressure within the system. Thus, the equilibrium will adjust to the left, forming more N_2O_4 molecules and reducing the number of molecules per unit volume. A suitable apparatus for the study of this effect is shown in Fig. 11.10.

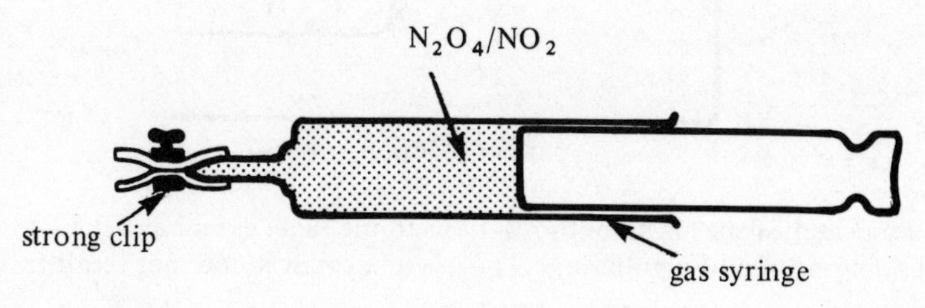

Fig. 11.10

The results are shown in the table below.

*Table 1 1.1

applied pressure change	initial colour change	final colour change
increase (plunger in)	darkens due to compression	lightens as equilibrium shifts to the left
decrease (plunger out)	lightens due to expansion	darkens as equilibrium shifts to the right

Generally in a reversible reaction involving a gas or gases at equilibrium, an increase in pressure will cause the equilibrium to shift in the direction which results in a decrease in the number of gaseous molecules. In a system in which there is no overall change in the total number of gaseous molecules changes in pressure will have no effect on the equilibrium position, e.g.

$$\underset{\text{1 mole}}{CO_{(g)}} + \underset{\text{1 mole}}{H_2O_{(g)}} \rightleftharpoons \underset{\text{1 mole}}{CO_{2\,(g)}} + \underset{\text{1 mole}}{H_{2\,(g)}}$$

Catalysts and Equilibrium

A catalyst speeds up a reaction by lowering the activation energy. However, in a reversible reaction it reduces the activation energy for both the forward reaction and the reverse reaction by the same amount, as shown in Fig.11.1

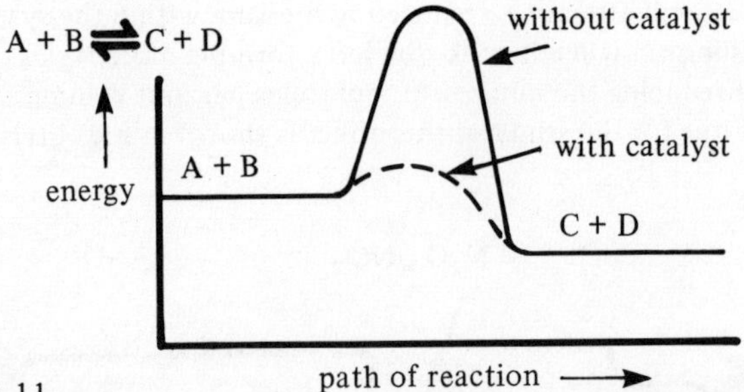

Fig. 11.11

Thus, a catalyst speeds up both reactions to the same extent and does *not* alter the position of equilibrium. The use of a catalyst does not result in an

*Authors' note – March 1980: Teachers may wish to consult *School Science Review* 1978, 211, **60**, 309 for an alternative explanation of the changes observed in the NO_2/N_2O_4 system.

increased yield of product. The advantage of using a catalyst in a reversible reaction is that it enables equilibrium to be reached more rapidly than without.

Summary

Table 11.2 summarises what we have considered in the previous two sections of this chapter.

Table 11.2

change applied	effect on equilibrium position
concentration:	
(1) addition of reactant or removal of product	equilibrium shifts to the right
(2) addition of product or removal of reactant	equilibrium shifts to the left
temperature:	
(1) increase	shifts in direction of endothermic reaction
(2) decrease	shifts in direction of exothermic reaction
pressure:	
(1) increase	shifts in direction which reduces the number of molecules in gas phase
(2) decrease	shifts in direction which increases the number of molecules in the gas phase
catalyst:	no effect on equilibrium position; equilibrium more rapidly attained

Industrial Processes

1. The Contact process:

$$\underbrace{2\,SO_{2\,(g)} + O_{2\,(g)}}_{3\text{ vol}} \rightleftharpoons \underbrace{2\,SO_{3\,(g)}}_{2\text{ vol}}$$

forward reaction is exothermic

We would expect the following conditions to favour the forward reaction and give a high yield of sulphur trioxide: low temperature and high pressure. However, at low temperatures the system takes a long time to reach equilibrium. In practice, a temperature of about 450°C is used, thus sacrificing maximum yield, but increasing the rate of reaction and a catalyst is used in order to speed up the attainment of equilibrium. It was found that the increase in yield due to high pressure was not sufficient to justify the heavy capital cost of the special equipment required.

Operating conditions in the Contact process: (1) 450°C, (2) atmospheric pressure, (3) catalyst – vanadium pentoxide pellets, V_2O_5.

2. The Haber process

$$\underbrace{N_{2\,(g)} + 3\,H_{2\,(g)}}_{4\text{ vol}} \rightleftharpoons \underset{2\text{ vol.}}{2\,NH_{3\,(g)}}$$

forward reaction is exothermic

Once again the conditions for maximum yield are: low temperature, high pressure. Fig. 11.12 shows the percentage of NH_3 in equilibrium when reacting nitrogen and hydrogen in a 1:3 mixture by volume at different temperatures and pressures.

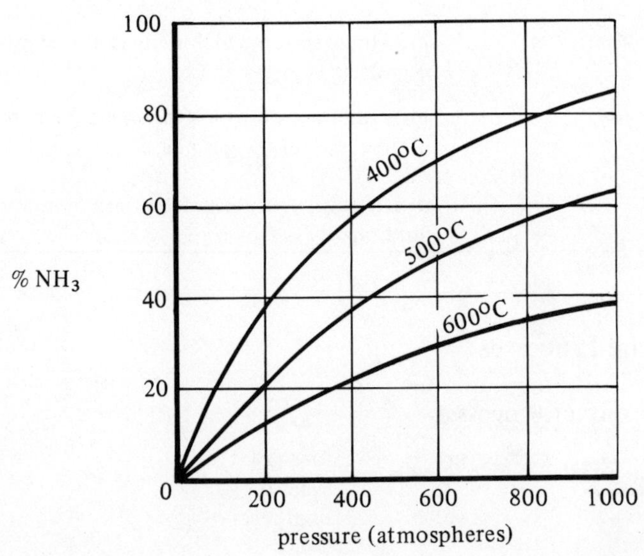

Fig. 11.12

The lower the temperature is, the higher the percentage of NH_3, but the slower the reaction. The higher the pressure the higher the percentage of ammonia, but the greater the cost of equipment both in outlay and maintenance.

Operating conditions: (1) 400 – 500°C, (2) high pressure, 200 – 1000 atmospheres, (3) iron catalyst. The ammonia is condensed on cooling and unreacted nitrogen and hydrogen are recycled.

Hydrolysis of Salts in Water

Pure water ionises to a very slight extent providing equal numbers of H^+ and OH^- ions.

$$H_2O \rightleftharpoons H^+_{(aq)} + OH^-_{(aq)}$$

The equilibrium lies far to the left as indicated by the heavy arrow in the above equation. The addition of a salt to water may upset this equilibrium.

(1) The solution of a salt formed from a strong base and a weak acid will have a pH greater than 7. **Examples**: sodium and potassium salts of ethanoic acid, carbonic acid and sulphurous acid. e.g. Sodium ethanoate in water:

Ions present initially: Na^+ + CH_3COO^- ← from the salt

$CH_3COOH \rightleftharpoons$ H^+ + OH^- ← from the water (a few)

Ethanoic acid is only partially dissociated; H^+ ions removed; water equilibrium shifts to the right; therefore, excess OH^- ions formed and $pH > 7$.

(2) The solution of a salt of a weak base and a strong acid will have a pH less than 7. **Examples**: chlorides, nitrates and sulphates of ammonium and metals *not* in groups I and II of the periodic table. e.g. Ammonium chloride in water:

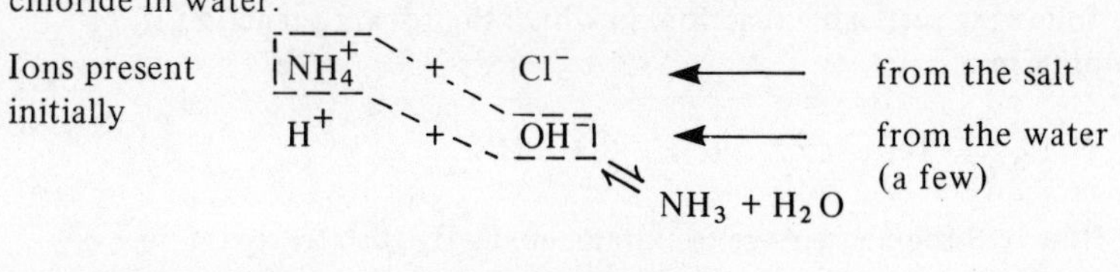

Ammonia solution is only partially dissociated; OH^- ions removed; water equilibrium shifts to the right; therefore, excess H^+ ions formed and $pH < 7$.

(3) The solution of a salt of a strong base and a strong acid will have a pH equal to 7. **Examples:** chlorides, nitrates and sulphates of sodium and potassium. e.g. Sodium chloride in water:

Ions present	Na^+	+	Cl^- ⟵	from the salt
	H^+	+	OH^- ⟵	from the water (a few)

Both acid and base are fully dissociated in solution; water equilibrium is not disturbed and hence pH = 7.

Examples for practice

1. When chlorine is dissolved in water the following equilibrium is set up.

$$Cl_2 + H_2O \rightleftharpoons 2\,H^+ + ClO^- + Cl^-$$

The hypochlorite ion, ClO^-, is responsible for the bleaching action of this solution. What effect on the bleaching efficiency of a solution of chlorine in water would the following have?

(a) Adding dilute nitric acid.
(b) Adding sodium chloride crystals.
(c) Adding potassium hydroxide solution.

2 Reaction (1): $H_{2\,(g)} + I_{2\,(g)} \rightleftharpoons 2\,HI_{(g)}$

Reaction (2): $2\,CO_{(g)} + O_{2\,(g)} \rightleftharpoons 2\,CO_{2\,(g)}$

Reaction (3): $CH_3OH_{(g)} \rightleftharpoons CO_{(g)} + 2\,H_{2\,(g)}$

(a) In which of the above reactions will an *increase* in pressure,
  (i) shift the position of equilibrium to the right,
  (ii) have no effect on the equilibrium position.
(b) In reaction (1), the forward reaction is exothermic. What effect, if any, will an increase in temperature have on the equilibrium position?

3. Butadiene, C_4H_6, an important intermediate in the manufacture of synthetic rubber, is produced from butene. One possible method involves the following reversible reaction, in which the forward reaction is endothermic.

$$C_4H_{8\,(g)} + I_{2\,(g)} \rightleftharpoons C_4H_{6\,(g)} + 2\,HI_{(g)}$$

(a) How will the percentage of butadiene, C_4H_6, be affected if an equilibrium mixture is subjected to an increase in
  (i) temperature, and (ii) pressure?

(b) Why does the presence of an alkali or a basic oxide have an effect on the equilibrium mixture?

4. The industrial preparation of methanol involves the combination of carbon monoxide and hydrogen using a zinc oxide catalyst according to the equation:

$$CO_{(g)} + 2\,H_{2\,(g)} \rightleftharpoons CH_3OH_{(g)}$$

The following graphs show the percentages of methanol in the reaction mixtures under different conditions.

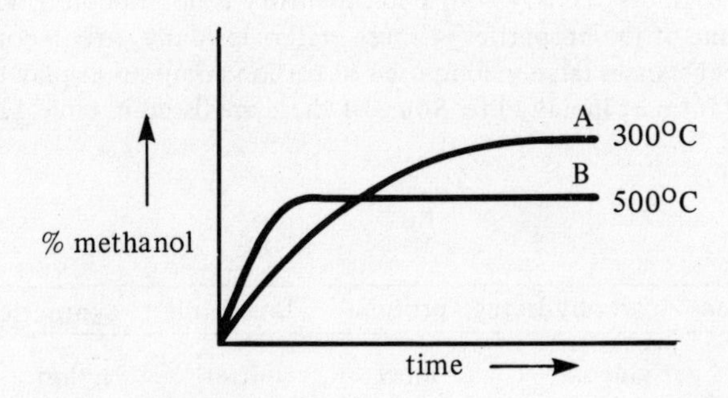

(a) Why does graph B slope more steeply than graph A at the start?
(b) Why do both graphs level off?
(c) Is the reaction which produces methanol exothermic or endothermic? Explain how you arrive at your answer.
(d) In industry, the reaction is usually carried out at 300 atmospheres pressure. Explain the use of high pressure.
(e) Copy graph B and using the same axes, draw a second graph to show how the percentage of methanol would have varied with time if no catalyst had been used.

(S.C.E.E.B.)

12 Introduction to the Chemistry of Carbon Compounds

Importance of Carbon Compounds

Compounds which contain carbon account for over 80% of all known chemical substances. Thus, a course in Chemistry is not complete without studying some of the properties of these materials. Many carbon compounds or substances largely composed of carbon compounds play a very important part in daily life. Some of these are listed in table 12.1 below.

Table 12.1

hydrocarbons	carbohydrates	proteins	fats & oils	synthetic polymers
calor gas	glucose	meat	butter	nylon
petrol	sugar	cheese	lard	polythene
kerosene	starch	egg white	suet	perspex
bitumen	cotton	wool	olive oil	polystyrene
rubber	cereals	silk	castor oil	PVC

In addition, other important materials can be derived from these substances, e.g.

1. glucose when fermented yields *ethanol* which on oxidation forms *ethanoic acid (acetic acid);*
2. cotton can be reconstituted to produce *rayon;*
3. vegetable oils can be hardened by reaction with hydrogen to form *margarine;*
4. fats and oils when treated with hot alkali produce *soap.*

Some Characteristic Features of Carbon and its Compounds

1. Chemical combinations

Of all the elements known, carbon has by far the greatest ability to combine with itself to form stable chain and ring molecules of varying size and complexity. This is largely due to the strength of bonds between carbon atoms. Table 12.2 shows the strength of single bonds between like atoms. The significant feature is the relatively high value for the carbon-carbon single bond.

Table 12.2

bond	bond energy (kJ mol^{-1})
C – C	347
Si – Si	176
N – N	163
O – O	146

2. Molecular mass.

Many carbon compounds have very high molecular masses. Naturally-occuring polymers such as proteins, polysaccharides and rubber, and synthetic polymers contain very large molecules, often referred to as **macromolecules**, and can have molecular masses well over 100 000.

3. Valency

In forming compounds carbon exhibits a valency of four, forming covalent bonds by sharing electrons with other atoms. Each carbon atom has a share of eight electrons and thereby achieves a complete outer shell.

4. Molecular properties

Most carbon compounds consist entirely of covalent molecules. As a result they *tend* to have the following properties:

(a) low boiling points and low melting points, due to there being only very weak van der Waals' forces *between* the molecules;

(b) non-conductors of electricity, unless they form ions in water e.g. acids and amines;

(c) immiscible with or insoluble in water, exceptions include compounds with –OH groups, e.g. ethanol, sucrose, which can form hydrogen bonds with water;

(d) miscible with or soluble in organic solvents such as benzene, ether, chloroform, acetone etc;

(e) reactions involving carbon compounds are usually slower than those in which ions are free in solution, since covalent bonds have to be broken before new bonds can be formed.

Homologous Series and Functional Groups

The chemistry of carbon compounds or **organic chemistry**, as it is otherwise known, is simplified by the fact that compounds can be grouped together in 'families' called **homologous series.** Members of a given series, known as homologues, usually show the characteristics listed below.
(1) Physical properties show a gradual change from one member to the next.
(2) Chemical properties are very similar.
(3) Homologues can be prepared by similar methods.
(4) The difference between successive members is $-CH_2-$ and consequently molecular masses differ by 14.
(5) They can be represented by a general formula. e.g. Alkanes C_xH_{2x+2}, Alkenes C_xH_{2x}
(6) Members contain the same **functional group.** By this is meant that the members contain a particular type of carbon-carbon bond and/or a certain group of atoms which is mainly responsible for the characteristic properties of that homologous series.

Table 12.3 examples of homologous series

name of series	functional group	example
alkenes	double bond, C=C	ethene, $CH_2{=}CH_2$
alcohols	hydroxyl, –OH	ethanol, C_2H_5OH
carboxylic acids	carboxyl, –COOH	ethanoic acid, CH_3COOH
amines	amine, $-NH_2$	ethylamine, $C_2H_5NH_2$

Classification of Carbon Compounds

Aliphatic: In these compounds the carbon atoms are linked together to form *chains*, e.g. ethene, octane, ethanol. Most of our study in the remaining chapters will be devoted to aliphatic compounds.

Alicyclic: In these compounds the carbon atoms are linked together to form *rings*, e.g. cyclohexane, cyclohexene.

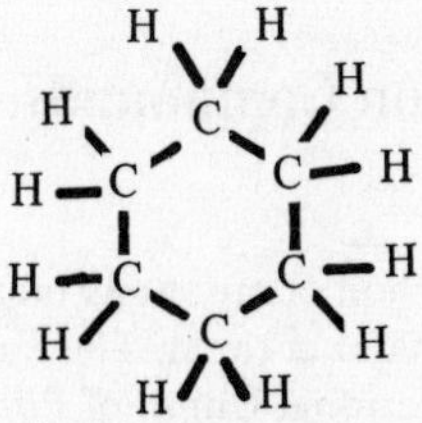

cyclohexane

Aromatic: These compounds also have a ring structure in which the molecules contain special benzene-type rings of six carbon atoms. Examples include benzene itself as well as toluene, phenol and aniline. We shall consider the structure of benzene in the next chapter.

Types of Formulae Used in Organic Chemistry

Empirical formula. This indicates the simplest ratio of each kind of atom present in the molecule. It is found by analysing the compound to find the percentage of each element present.

Molecular formula. This indicates how many atoms of each element are present in the molecule. It can be determined once the empirical formula and the molecular mass are known.

Structural formula. This indicates how the atoms are arranged in the molecule. This can be shown in (a) an **extended** form in which all the bonds are indicated or in (b) a **condensed** or abbreviated form where atoms are grouped together.

Table 12.4 examples of different formulae

compound	ethane	ethanol	ethanoic acid
empirical formula	CH_3	C_2H_6O	CH_2O
molecular formula	C_2H_6	C_2H_6O or C_2H_5OH	$C_2H_4O_2$ or CH_3COOH
structural formula (a) extended	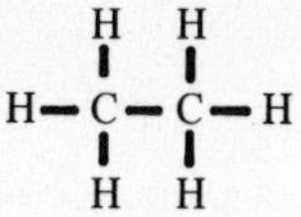 H H H—C—C—H H H	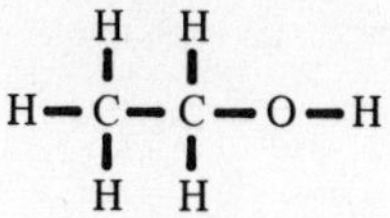 H H H—C—C—O—H H H	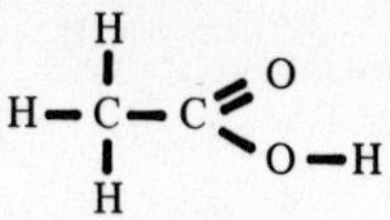 H O H—C—C H O—H
(b) condensed	CH_3CH_3	CH_3CH_2OH	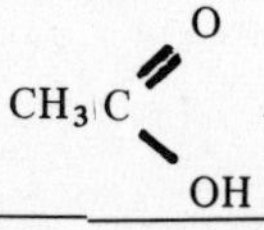 CH_3C (=O) (—OH)

Nomenclature of Carbon Compounds

As the number of known carbon compounds increased so the need grew for a systematic method of naming them. The method currently in use is that laid down by the International Union of Pure and Applied Chemistry, I.U.P.A.C. for short. We shall see how the method is applied to different homologous series in subsequent chapters.

Examples for practice. These are to be found at the end of Chapter 17.

13 Hydrocarbons (1): their Molecular Structure

This chapter deals with the way in which atoms are arranged within the hydrocarbon molecules. The study of molecular structure is known as stereochemistry.

Classification of Hydrocarbons

Hydrocarbons can firstly be classified into the three main groups mentioned in the previous chapter, namely **aliphatic, alicyclic and aromatic.** The aliphatic hydrocarbons can be further subdivided.

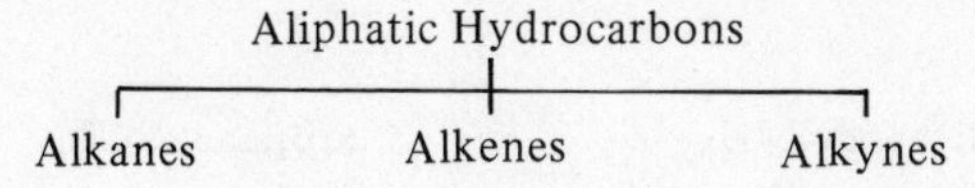

Table 13.1 summarises some of the important features and gives examples of the three aliphatic series which will be covered in this chapter and the next.

Table 13.1

	alkanes	alkenes	alkynes
characteristic bond types	all single bonds i.e. C—C and C—H	one carbon-carbon double bond, C=C	one carbon-carbon triple bond, C≡C
name ending	– ANE	– ENE	– YNE
general formula	C_xH_{2x+2}	C_xH_{2x}	C_xH_{2x-2}
examples	methane, CH_4		
	ethane, C_2H_6	ethene, C_2H_4	ethyne, C_2H_2
	propane, C_3H_8	propene, C_3H_6	propyne, C_3H_4
	butane, C_4H_{10}	butene, C_4H_8	butyne, C_4H_6
	pentane, C_5H_{12}	pentene, C_5H_{10}	pentyne, C_5H_8
	hexane, C_6H_{14}	hexene, C_6H_{12}	hexyne, C_6H_{10}

Data

The tables of data given below are relevant to the topic of molecular structure

Table 13.2 data relating to types of bonds

bond	bond energy (kJ mol^{-1})	bond length (Å, 10^{-10} m)
C–C (aliphatic)	347	1.54
C=C	598	1.34
C≡C	811	1.20
C⋯C (aromatic)	519	1.39

Table 13.3 heats of formation

compound	ΔH_f (kJ mol^{-1})
ethane, C_2H_6	– 85
ethene, C_2H_4	+ 52
ethyne, C_2H_2	+ 227
benzene, C_6H_6	+ 49

Alkanes

Methane, CH_4 Ethane, C_2H_6 Propane, C_3H_8

```
   H          H  H                 H  H  H
   |          |  |                 |  |  |
H— C — H   H— C— C —H ; CH3CH3  H— C— C— C —H ; CH3CH2C
   |          |  |                 |  |  |
   H          H  H                 H  H  H
```

The next homologue is butane, C_4H_{10}, and it can have two possible structures, as follows –

```
                                   H
                                   |
(1) H  H  H  H             (2)  H— C —H
    |  |  |  |                     |
 H— C— C— C— C —H              H   |   H                CH3
    |  |  |  |                 |   |   |                 |
    H  H  H  H              H— C — C — C —H  ;      CH3CHCH3
                               |   |   |
    CH3CH2CH2CH3               H   H   H
```

These compounds have the same number of atoms of carbon and hydrogen, i.e. the same molecular formula, but they have different structural formulae. Such compounds are called **isomers.**

Compound (1) is called normal butane, abbreviated to n-butane, and is an example of a straight-chain hydrocarbon. Compound (2) is a branched

hydrocarbon and its name is 2-methylpropane. The method of naming this compound follows a system laid down by the International Union of Pure and Applied Chemistry (I.U.P.A.C.) and operates as follows:
(1) Select the longest chain of carbon atoms in the molecule and name it after the appropriate compound.
(2) Find out which atom the branch is attached to by giving each carbon in the longest chain a number. Begin numbering from the end which is nearer the branch.
(3) Identify the branch, i.e. whether it is a methyl group (CH_3-) or ethyl group (C_2H_5-) etc.

Fig. 13.1 illustrates how this system applies to the above example.

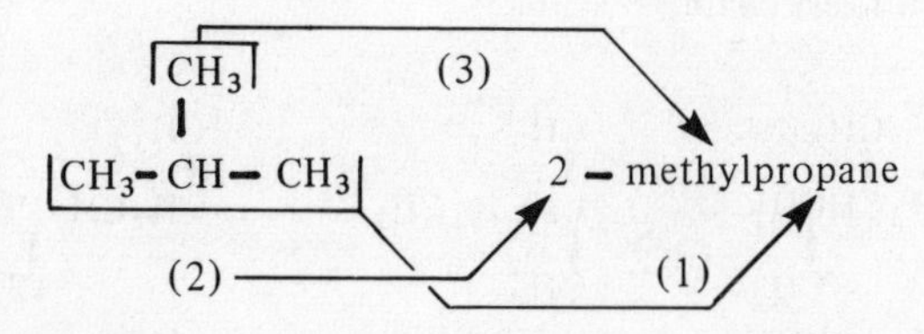

Fig. 13.1

The two isomers of butane can be distinguished in a number of ways. They can be separated by gas chromatography and they differ in boiling point; n-butane boils at 0°C, 2-methylpropane at −10°C.

Pentane has three isomers as follows: (boiling points given in brackets)

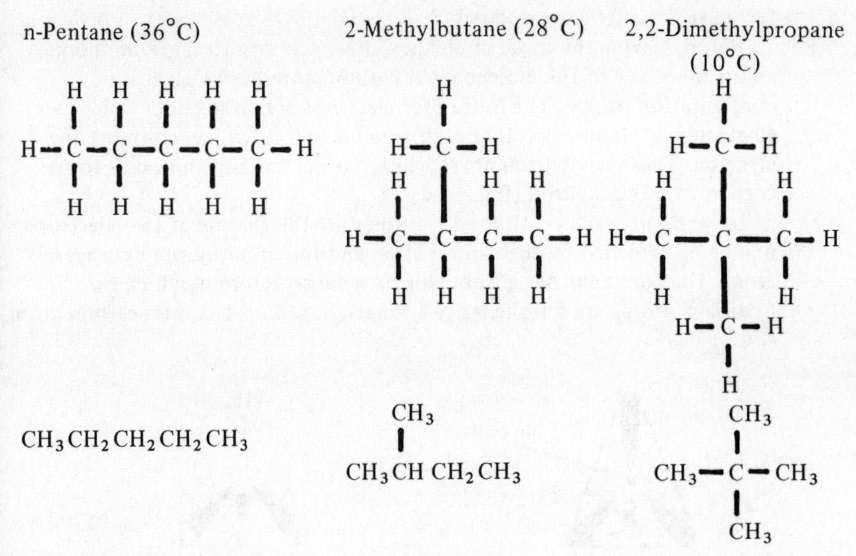

There is a tendency for the boiling point to decrease as the amount of branching increases, since this causes the molecule to become more compact and have a smaller surface area. The weak forces of attraction between the molecules are thus further reduced. The graph of melting and boiling points shown in the next chapter uses the values for the straight-chain alkanes only.

The following structures are given to illustrate some of the problems which arise when dealing with structural formulae. Additional isomers are *not* obtained by

(1) writing the branch below instead of above the carbon atom chain, or
(2) numbering from the wrong end, or
(3) having a bend in the longest chain.

$$\begin{array}{l} CH_3CH_2\underset{\displaystyle CH_3}{\underset{|}{CH}}CH_3 \\ \text{(a)} \end{array} \qquad \begin{array}{l} CH_3\overset{\displaystyle CH_3}{\overset{|}{CH}}\underset{\displaystyle CH_3}{\underset{|}{CH_2}} \\ \text{(b)} \end{array} \qquad \begin{array}{l} \underset{\displaystyle CH_3}{\underset{|}{\overset{\displaystyle CH_3}{\overset{|}{CH}}}}CH_2CH_3 \\ \text{(c)} \end{array} \qquad \begin{array}{l} \overset{\displaystyle CH_3}{\overset{|}{CH_2}}CH_2\underset{\displaystyle CH_3}{\underset{|}{CH_2}} \\ \text{(d)} \end{array}$$

Structure (d) is n-pentane; (a), (b) and (c) are all other ways of representing 2-methylbutane!

Isomerism is a further reason for the large number of known carbon compounds. Decane, $C_{10}H_{22}$, has 75 isomers. The theoretical number of isomers of tricosane, $C_{30}H_{62}$ exceeds 4 000 000 000, although only very few have actually been isolated.

Before leaving the topic of alkanes, it is necessary to say something about the shape of the molecules. A carbon atom has an electronic configuration of 2,4. The four outer electrons are arranged in four electron-pair clouds such that each cloud is half-filled, i.e. contains one electron. These compartments repel each other equally and point to the corners of a tetrahedron (Fig. 13.2).

In methane, each C–H bond is formed by the sharing of two electrons, one being provided by the carbon atom and the other by the hydrogen atom. Thus methane has a three-dimensional structure in which the hydrogen atoms are tetrahedrally arranged around the central carbon atom (Fig. 13.3).

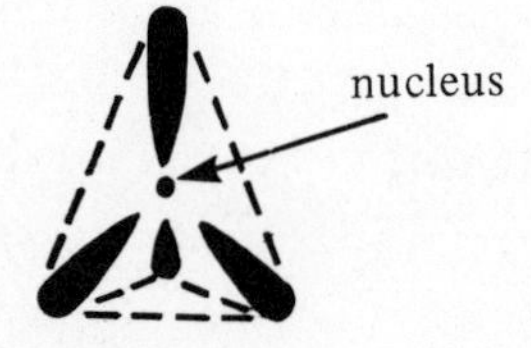

Fig. 13.2 carbon atom

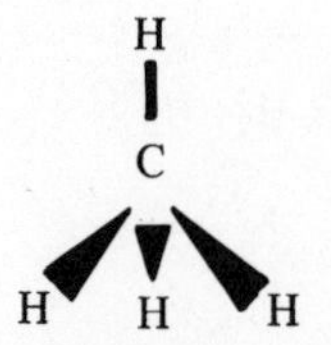

Fig. 13.3 methane

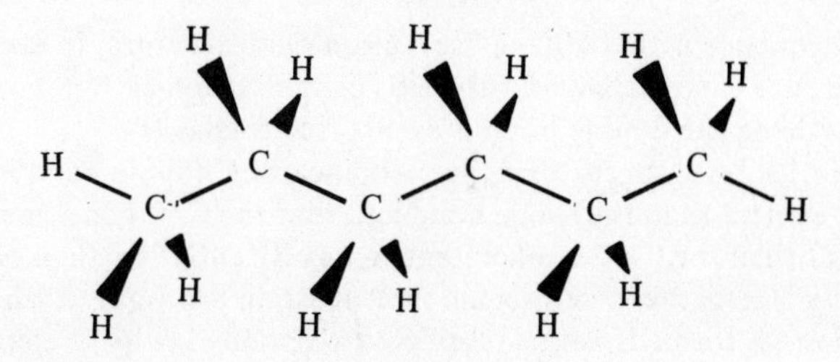

Fig. 13.4 hexane

The angle between any pair of C–H bonds in methane is the same (just less than 110°). This tetrahedral arrangement also occurs for each carbon atom in a longer chain alkane, e.g. hexane (Fig. 13.4). This zig-zag structure is not rigid.

Alkenes

In a molecule of ethene, each carbon atom has two hydrogen atoms covalently bonded to it. The double bond between the carbon atoms is formed by the overlap of two half-filled electron pair clouds from each carbon atom. (Fig. 13.5).

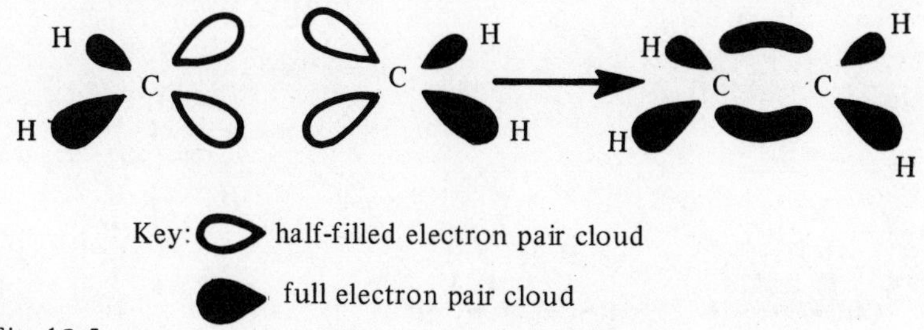

Fig. 13.5

As a result, the molecule of ethene has a planar structure (Fig. 13.6) in contrast to the three-dimensional shape of ethane (Fig. 13.7).

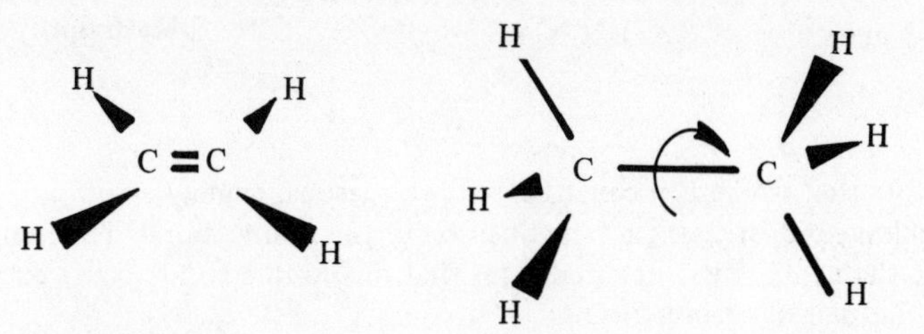

Fig. 13.6 ethene

Fig. 13.7 ethane

Another consequence is that ethene has a more rigid structure. In ethane, the methyl groups can rotate about the axis of the carbon-carbon bond. However, in ethene the double bond prevents free rotation.

From table 13.2 we can see that the carbon-carbon double bond is considerably shorter than the single bond and that the C=C bond energy, though greater than the C–C bond energy, is significantly less than twice as great. This suggests that a certain amount of strain is produced when the double bond is formed, as this results in two bonded pairs of electrons being brought close together. The slightly endothermic heat of formation of ethene (table 13.3) suggests further evidence for this. When alkenes undergo addition across the double bond, this strain is relieved as the carbon atoms involved are now held together by a single bond.

Structural formulae

Ethene and propene each possess only one structure, but butene has a number of isomers.

Ethene, C_2H_4

```
H     H
 \   /
  C=C     ; CH2=CH2
 /   \
H     H
```

Propene, C_3H_6

```
   H   H
   |  /
 H-C-C=C   ; CH3CH=CH2
   | |  \
   H H   H
```

But-1-ene

```
H    H H H
 \   | | |
  C=C-C-C-H
 /    | |
H     H H
```

$CH_2{=}CHCH_2CH_3$

But-1-ene

But-2-ene

```
  H H H H
  | | | |
H-C-C=C-C-H
  |     |
  H     H
```

$CH_3CH{=}CHCH_3$

But-2-ene

2-Methylpropene

```
       H
       |
     H-C-H
H      |  H
 \     |  |
  C=C-C-H
 /      |
H       H
```

$$\begin{array}{c} \quad\quad CH_3 \\ \quad\quad | \\ CH_2 = CCH_3 \end{array}$$

2-Methylpropene

Note that isomerism can arise for two reasons, namely chain branching and/or changing the position of the double bond. The number '1' in the name 'but-1-ene' indicates that the double bond occurs between the first pair of carbon atoms.

The rigidity of structure caused by the double bond means that there are two possible structures of but-2-ene as shown below.

trans-But-2-ene (methyl groups on *opposite* sides of the double bond)

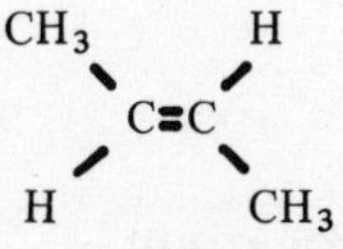

$$\begin{matrix} CH_3 & & H \\ & C{=}C & \\ H & & CH_3 \end{matrix}$$

cis-But-2-ene (methyl groups on the *same* side of the double bond)

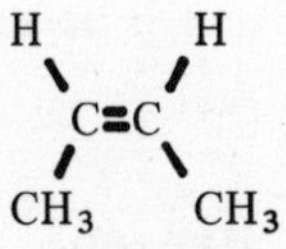

$$\begin{matrix} H & & H \\ & C{=}C & \\ CH_3 & & CH_3 \end{matrix}$$

Alkynes

In ethyne, each carbon atom has one hydrogen atom covalently bonded to it and the carbon atoms are linked together by the overlap of three half-filled electron-pair clouds from each atom. Since all four atoms lie in a straight line, the ethyne molecule is said to have a linear structure.

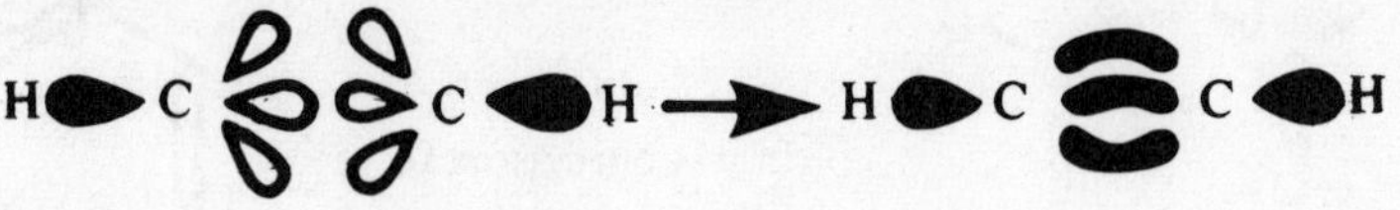

H C C H → H C C H

Fig. 13.8

In the ethyne molecule we would expect an even greater strain between the three bonded pairs of electrons which hold the two carbon atoms together. As an indication of this it is worth noting the following points from tables 13.2 and 13.3.

(1) The C≡C bond length is much shorter than those of the C–C or C=C.
(2) The C≡C bond energy is much less than three times the C–C bond energy.
(3) The heat of formation of ethyne is highly endothermic.

Structural formulae

Ethyne, C_2H_2 Propyne, C_3H_4 Butyne, C_4H_6, has two isomers.

H–C≡C–H

CH≡CH

$$H{-}C{\equiv}C{-}\overset{\displaystyle H}{\underset{\displaystyle H}{C}}{-}H$$

CH≡C CH_3

CH≡C CH_2CH_3

But-1-yne

CH_3C≡C CH_3

But-2-yne

It is only fair to point out that the explanation given above for the formation of the carbon-carbon double and triple bonds is oversimplified. The full explanation is considered to be too complex to be dealt with at this stage. Should you decide to continue your study of Chemistry beyond this year you will very likely come across a fuller and more accurate description of the carbon-carbon bonding.

Benzene – an Aromatic Hydrocarbon

The molecular formula of benzene, C_6H_6, suggests the likelihood that (a) it is unsaturated and (b) it has a ring structure. Over the years many attempts have been made to unravel the secret of benzene's structure. One of the most important of these is illustrated below in which the carbon atoms are linked together in a ring by means of alternate single and double bonds.

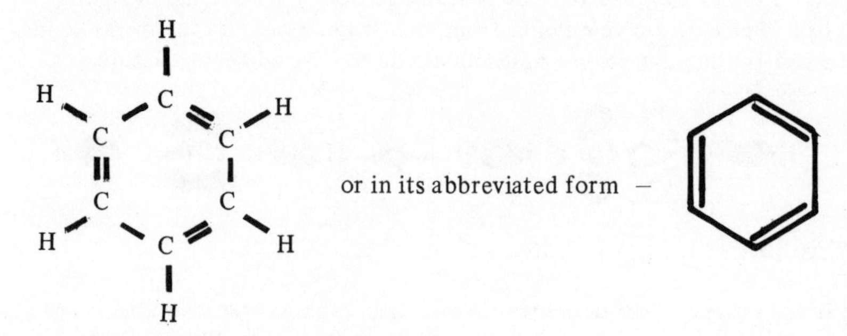

Benzene does *not* decolourise bromine water, however, thus indicating the absence of double bonds.

Other experimental investigations into the structure of benzene have shown that

(1) its molecule is planar, i.e. all carbon and hydrogen atoms lie in the same plane, and

(2) the ring of six carbon atoms forms a regular hexagon, i.e. all the carbon-to-carbon bond lengths are the same.

If the bonding in benzene were alternate single and double bonds, the hexagon would not be regular, since double bonds are shorter than single. This is shown in exaggerated form here.

In benzene, each carbon atom uses up three of its four outer electrons in forming bonds with a hydrogen atom and two adjacent carbon atoms. The six remaining outer electrons, one from each carbon atom, occupy electron clouds which are not confined or localised between any one pair of carbon atoms. These electrons are said to be **delocalised** (Fig. 13.9).

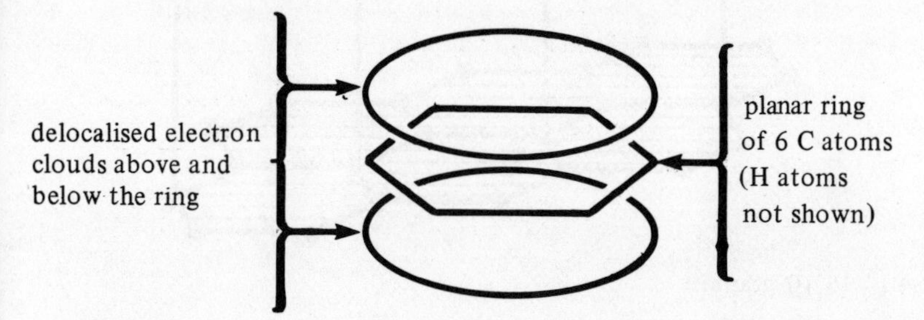

Fig. 13.9

The delocalised electrons give additional bonding strength, so that the carbon-carbon bond energy in benzene is in between that of the aliphatic C–C and C=C bond energies.
Note: from table 13.2 that the bond length is also intermediate in value.
The structure of benzene is usually represented nowadays as follows:

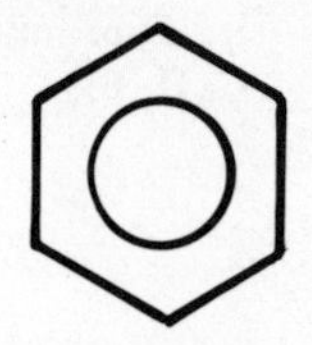

Each corner of the hexagon indicates the position of a carbon atom and hydrogen atoms are not shown. The circle inside the hexagon indicates the additional bonding due to the delocalised electrons.

It is interesting to compare this structure with that of graphite, one of the polymorphs of carbon, in which the atoms are arranged in layers (Fig. 13.10). Within each layer, the carbon atoms are linked in regular six-membered rings giving a 'honeycomb' appearance. In graphite, the 'spare' electrons are so much more delocalised that they can be made to flow from one ring to another and thus graphite can conduct an electric current in a similar manner to a metal.

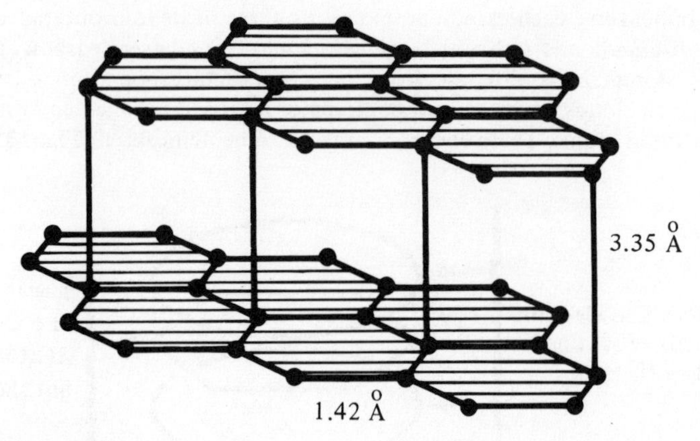

Fig. 13.10 graphite

The structural formulae of some simple aromatic compounds are shown below.

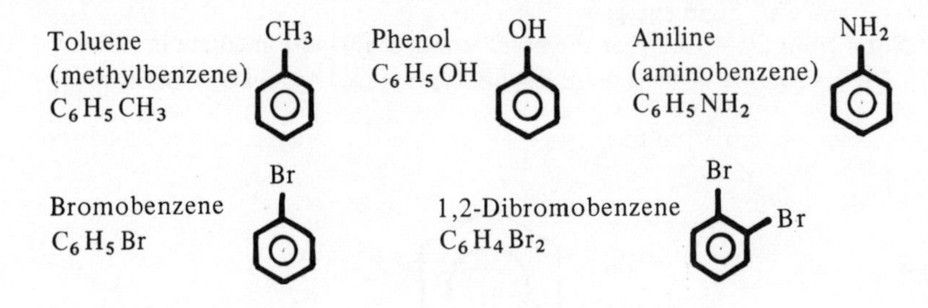

Alicyclic Hydrocarbons

These compounds bear a closer resemblance to aliphatic hydrocarbons in structure and chemical behaviour than to aromatic. Cyclohexane, C_6H_{12}, has a ring structure which is puckered as shown in Fig. 13.11. It is also referred to as a 'chair' structure. As far as chemical properties are concerned cyclohexane is similar to an alkane and cyclohexene to an alkene.

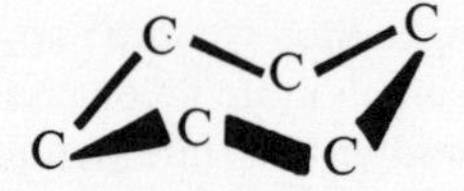

Fig. 13.11 puckered ring of 6 C atoms (H atoms not shown)

Examples for practice

1. Write the structural formulae and systematic names of the 5 isomers of hexane.

2. Write the structural formulae of
(a) 2,4-dimethylhexane,
(b) 3-ethyl-3-methylpentane.

3. Write the systematic names for

$$\text{(a)}\quad \begin{array}{c}\;CH_3\qquad\quad CH_3\\ \;|\qquad\qquad\;\; |\\ CH_3CHCH_2CHCHCH_3\\ \qquad\qquad\quad |\\ \qquad\qquad\quad C_2H_5\end{array}\qquad\qquad \text{(b)}\quad \begin{array}{c}\qquad\;\; CH_3\qquad CH_3\\ \qquad\;\; |\qquad\qquad |\\ CH_3-C-CH_2CHCH_3\\ \qquad\;\; |\qquad\qquad\quad\\ \qquad\;\; CH_3\qquad\qquad\end{array}$$

The structure shown in question 3(b) is also known as iso-octane and is the standard by which fuels are given an octane rating.

4. Write the structural formula of (a) 3-ethylpent-2-ene, (b) 2,3-dimethylbut-2-ene.

5. Give the structural formulae and systematic names of the 3 isomers of pentyne.

6. Give the structural formulae and systematic names of (a) the 2 isomers of dibromoethane, (b) the 3 isomers of dibromoethene and (c) the 3 isomers of dibromobenzene.

14 Hydrocarbons (2): How they React

Sources of Hydrocarbons

1. Natural gas

Natural gas fields are exploited commercially in many parts of the world, notably in Libya, U.S.A., Italy and Holland as well as under the North Sea. The main constituent is methane, but natural gas may also contain small amounts of ethane, propane and butane.

2. Petroleum or crude oil

Petroleum is a complex mixture of hydrocarbons. Although it is a liquid, it contains dissolved gases and solids. At an oil refinery, crude oil is first distilled to yield a series of boiling-point ranges called **fractions.** Since much of crude oil consists of long-chain hydrocarbons, for which there is a limited demand, some of the higher boiling fractions are subjected to a second process known as **cracking** in which the long-chain molecules are broken down to give simpler and more useful compounds. The process is usually carried out at high temperatures and in the presence of a catalyst – a special form of aluminium silicate. Ethene and other unsaturated hydrocarbons are also formed during cracking.

Although recent years have seen the discovery of new oil fields in Alaska and under the North Sea, the Middle East promises to be the main oil supplier for a long time yet. The overall composition of crude oil varies from one region to another. For example, Middle East oil consists mainly of alkanes, while oil from California and Mexico is relatively rich in aromatic hydrocarbons and Venezuelan oil in alicyclic compounds.

3. Coal tar

When coal is heated in the absence of air, it is destructively distilled to yield coal gas, ammonia solution, coal tar and coke. Unlike crude oil, coal tar consists predominantly of aromatic compounds. These are separated by fractional distillation to give the starting materials for the production of dyes, plastics and drugs.

Alkanes

Physical properties

no. of C atoms per molecule:	1 – 4	5 – 17	> 17
state at room temperature:	gases	liquids	waxy solids

In chapter 12 it was mentioned that members of the same homologous series tend to show a gradation in physical properties. This is clearly shown in Fig. 14.1, particularly by the boiling points. The forces of attraction between the molecules increase as the chain length increases. The melting points do not fall on a smooth curve, but the general trend is similar.

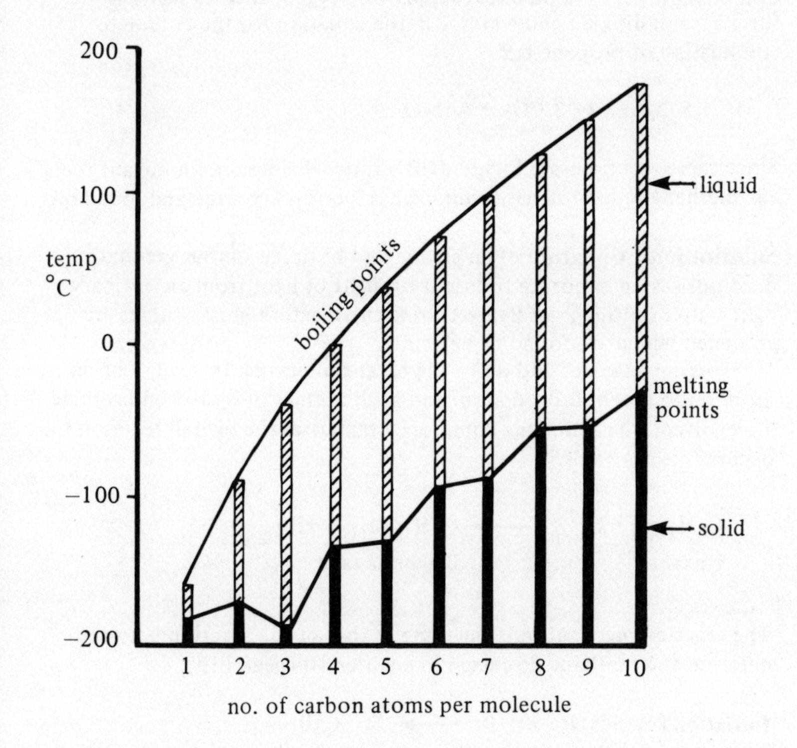

Fig. 14.1: boiling points & melting points of straight-chain alkanes.

Chemical properties

In many older text-books you may find that the alkanes are referred to as the paraffins. The word 'paraffin' means literally 'having little affinity' towards other substances or, in a word, 'unreactive'. To anyone who has seen petrol catch fire or a paraffin stove blaze up, the term 'unreactive' would appear to be hardly appropriate! However, alkanes are unreactive towards virtually all ionic reagents, such as acids, alkalis, oxidising agents and reducing agents.

Alkanes do not decolourise bromine water or acidified potassium permanganate solution. These results indicate that alkanes are **saturated,** i.e. they have only single bonds between carbon atoms and they contain the maximum possible number of hydrogen atoms. Alkanes do not, therefore, undergo addition reactions. However, they can react by **substitution** in which a hydrogen atom is replaced by a different atom.

Combustion: Given a plentiful supply of oxygen, alkanes will burn to form carbon dioxide and water, e.g. the equation for the complete combustion of propane is

$$C_3H_8 + 5\ O_2 \longrightarrow 3\ CO_2 + 4\ H_2O$$

Since these reactions are highly exothermic, alkanes are important fuels, e.g. methane in natural gas, camping gas, petrol, kerosene and diesel oil.

Substitution: A mixture of an alkane and bromine vapour gradually decolourises on exposure to bright sunlight or light from an ordinary light-bulb or ultra-violet lamp. During the reaction, misty fumes are produced which are found to be acidic.

Bromine does not **add** on to the alkane molecule. Instead, a bromine atom replaces a hydrogen atom and acidic fumes of hydrogen bromide are evolved. The following equation summarises the overall reaction that occurs.

$$\underset{\text{Hexane}}{C_6H_{14\,(l)}} + Br_{2\,(g)} \longrightarrow \underset{\text{Bromohexane}}{C_6H_{13}Br_{(l)}} + HBr_{(g)}$$

The reaction mechanism is the same as that of the reaction between chlorine and methane discussed in chapter 10 (page 118).

Initiation (by light): $\underset{\text{molecules}}{Br{-}Br} \longrightarrow Br^{\bullet} + {}^{\bullet}Br$

separate atoms (free radicals) which have unpaired electrons

Propagation (illustrated here with a simpler alkane than hexane, since this is a general reaction of alkanes):

$$(1)\quad \begin{array}{ccccc} & H & & H & \\ & | & & | & \\ H- & C & - & C & -H \\ & | & & | & \\ & H & & H & \end{array} + {}^{\bullet}Br \longrightarrow \begin{array}{cccc} & H & & H \\ & | & & | \\ H- & C & - & C^{\bullet} \\ & | & & | \\ & H & & H \end{array} + H-Br$$

$$(2)\quad \begin{array}{cccc} & H & & H \\ & | & & | \\ H- & C & - & C^{\bullet} \\ & | & & | \\ & H & & H \end{array} + Br-Br \longrightarrow \begin{array}{ccccc} & H & & H & \\ & | & & | & \\ H- & C & - & C & -Br \\ & | & & | & \\ & H & & H & \end{array} + {}^{\bullet}Br$$

The propagation steps are repeated many times. Substitution is not necessarily confined to one hydrogen atom per molecule or the same hydrogen. Consequently, a large variety of products is possible.

Chlorine and fluorine react similarly, although more vigorously than bromine. With iodine there is virtually no reaction. The products of this reaction belong to a series of compounds known as the **alkyl halides.**

Alkyl Halides

Alkyl halides are much more versatile chemically than alkanes because of the polarity of the carbon-halogen bond. This renders the carbon more prone to attack by other chemical species. Thus alkyl halides can undergo many different **substitution** reactions. We shall consider only one example at this stage, namely the replacement of the halogen atom by an -OH group to form an alcohol. For example:

$$\begin{array}{ccccccccc} & H & & H & & H & & H & \\ & | & & | & & | & & | & \\ H- & C & - & C & - & C & - & C & -Br \\ & | & & | & & | & & | & \\ & H & & H & & H & & H & \end{array} + OH^- \longrightarrow \begin{array}{ccccccccc} & H & & H & & H & & H & \\ & | & & | & & | & & | & \\ H- & C & - & C & - & C & - & C & -OH \\ & | & & | & & | & & | & \\ & H & & H & & H & & H & \end{array} + Br^-$$

1-Bromobutane   Butan-1-ol

This reaction can be represented more generally as follows:

$$R-X + OH^- \longrightarrow R-OH + X^-$$

where 'R' represents an alkyl group (e.g. methyl, ethyl etc) and '*X*' a halogen atom. In this reaction halide ions are released and can be detected by their formation of a precipitate with silver nitrate solution. When a few drops of silver nitrate solution are added to either iodobutane or bromobutane, a cloudiness appears due to the formation of silver halide precipitates. This occurs because the hydroxyl ions in the silver nitrate solution are able to release a sufficient number of halide ions.

However, no reaction appears to take place with 1-chlorobutane and a higher concentration of hydroxyl ions is required. One of the reasons for this lack of reaction is that the C–Cl bond is much stronger than the C–I or C–Br bonds (see page 202). In this case the alkyl halide is warmed in a solution of potassium hydroxide in alcohol for a few minutes, cooled and then acidified with dilute nitric acid (why this acid?). When silver nitrate solution is added a white precipitate is formed which shows that chloride ions have been released.

The substitution reactions mentioned in this and the previous sections illustrate that alkyl halides are useful intermediates between alkanes and alcohols. This can be summarised as follows:

Alkanes, R – H ⟶ Alkyl halides, R – *X* ⟶ Alcohols, R – OH

```
   H  H                  H  H                   H  H
   |  |                  |  |                   |  |
H––C––C––H   ⟶     H––C––C––Br     ⟶     H––C––C––OH
   |  |                  |  |                   |  |
   H  H                  H  H                   H  H
```

Ethane   Bromoethane   Ethanol

Alkenes

Making alkenes

Alkenes can be prepared in the laboratory by the **dehydration** of the appropriate alcohol. For example, ethanol yields ethene as shown in the following equation.

```
    H   H            H         H
    |   |             \       /
H – C – C – H   ⟶      C ═ C        + H2O
    |   |             /       \
   [H   OH]          H         H
```

Dehydration can be achieved by:

(1) heating a mixture of ethanol with excess concentrated sulphuric acid, or by

(2) a catalytic method. When ethanol vapour is passed over hot aluminium oxide it is catalytically dehydrated to give ethene.

Industrially, alkenes are produced during the catalytic cracking of long-chain hydrocarbons obtained from petroleum.

Chemical properties

Combustion: Like alkanes, alkenes burn in a plentiful supply of oxygen to form carbon dioxide and water.

Addition reactions: Alkenes readily undergo addition reactions to the carbon-carbon double bond. As a result, alkenes are said to be **unsaturated.** The tests for unsaturation are (1) decolourisation of bromine water, and (2) decolourisation of acidified potassium permanganate solution.

Halogens: Bromine, for example, adds across the double bond to form a colourless product.

$$\begin{array}{ccccccc} & H & & H & & & H \\ & | & & | & & / & \\ H- & C & - & C & = & C & \\ & | & & & & \backslash & \\ & H & & & & & H \end{array} \quad + Br_2 \longrightarrow \quad \begin{array}{ccccccc} & H & & H & & H & \\ & | & & | & & | & \\ H- & C & - & C & - & C & -H \\ & | & & | & & | & \\ & H & & Br & & Br & \end{array}$$

Propene 1,2-Dibromopropane

Hydrogen halides: These react similarly to form alkyl halides.

$$\begin{array}{ccccccc} & H & & H & & & H \\ & | & & | & & / & \\ H- & C & - & C & = & C & \\ & | & & & & \backslash & \\ & H & & & & & H \end{array} \quad + HBr \longrightarrow \quad \begin{array}{ccccccc} & H & & H & & H & \\ & | & & | & & | & \\ H- & C & - & C & - & C & -H \\ & | & & | & & | & \\ & H & & Br & & H & \end{array}$$

Propene 2-Bromopropane

Hydrogen:

$$\begin{array}{ccccc} H & & & & H \\ & \backslash & & / & \\ & C & = & C & \\ & / & & \backslash & \\ H & & & & H \end{array} \quad + H_2 \longrightarrow \quad \begin{array}{ccccc} & H & & H & \\ & | & & | & \\ H- & C & - & C & -H \\ & | & & | & \\ & H & & H & \end{array}$$

Ethene Ethane

In order to bring about this reaction, the reactants are mixed and heated in the presence of a catalyst, usually a special form of finely-divided nickel.

Hydrogenation, as this reaction is known, has important industrial application in the conversion of unsaturated vegetable oils into solid fats to make margarine. The greater the degree of hydrogenation, the harder is the product. The reaction can be depicted as follows.

$$\sim\!\sim\,\mathrm{CH{=}CH}\,\sim\!\sim \; + H_2 \longrightarrow \; \sim\!\sim\,\mathrm{CH_2{-}CH_2}\,\sim\!\sim$$

unsaturated molecule of oil — saturated molecule of fat

Sulphuric acid: Ethene reacts with cold concentrated sulphuric acid to form an addition product called ethyl hydrogen sulphate. On hydrolysis with water, ethanol is formed. The overall process is called **hydration** since it effectively involves the addition of water across the double bond.

$$\mathrm{H_2C{=}CH_2} \xrightarrow[\mathrm{H_2SO_4}]{\text{conc.}} \mathrm{H{-}CH_2{-}CH_2{-}O{-}SO_3H} \xrightarrow{H_2O} \mathrm{H{-}CH_2{-}CH_2{-}OH}$$

Ethene — Ethyl hydrogen sulphate — Ethanol

This reaction is also important industrially as it is used to remove alkenes during the manufacture of petrol. If they were not removed, they might polymerise forming undesirable products in the car engine.

Potassium permanganate: From its position on the table of E° values we can see that the permanganate ion in an acidified solution is a powerful oxidising agent or electron acceptor (page 203). Alkenes are oxidised by the addition of two -OH groups to the double bond.

$$\mathrm{H_2C{=}CH_2} \longrightarrow \mathrm{HO{-}CH_2{-}CH_2{-}OH}$$

Ethene, for example, yields a substance called ethan-1,2-diol, more commonly known as ethylene glycol, which is an important antifreeze.

Addition polymerisation: The most important use of alkenes is in the production of long-chain molecules called polymers. The general reaction can be represented as follows:

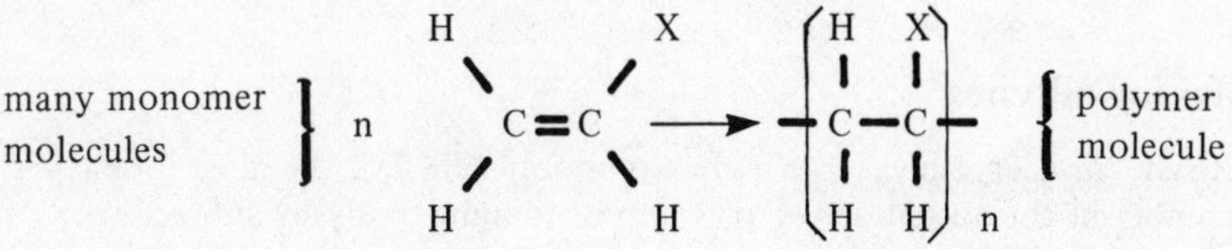

$$\text{many monomer molecules}\ \Big\}\ n\ \begin{matrix} H & & X \\ & C=C & \\ H & & H \end{matrix} \longrightarrow \left(\begin{matrix} H & X \\ | & | \\ -C & -C- \\ | & | \\ H & H \end{matrix} \right)_n \ \Big\{\ \text{polymer molecule}$$

name of polymer	polythene	polyvinyl-chloride(PVC)	polypropylene	polystyrene
side-chain, X–	H–	Cl–	CH_3–	C_6H_5–, ⌬–

The reaction frequently has to be initiated by a substance which will provide free radicals. Organic peroxides are suitable compounds since they possess relatively weak oxygen-oxygen bonds (page 137).

Initiation: $Y - Y \longrightarrow Y^{\bullet} + {}^{\bullet}Y$
initiator — free radicals having unpaired electrons

Propagation:

(1)
$$Y^{\bullet} + \begin{matrix} H & & X \\ & C=C & \\ H & & H \end{matrix} \longrightarrow Y-\begin{matrix} H \\ | \\ C \\ | \\ H \end{matrix}-\begin{matrix} X \\ | \\ C^{\bullet} \\ | \\ H \end{matrix}$$

(2)
$$Y-\begin{matrix} H \\ | \\ C \\ | \\ H \end{matrix}-\begin{matrix} X \\ | \\ C^{\bullet} \\ | \\ H \end{matrix} + \begin{matrix} H & & X \\ & C=C & \\ H & & H \end{matrix} \longrightarrow Y-\begin{matrix} H \\ | \\ C \\ | \\ H \end{matrix}-\begin{matrix} X \\ | \\ C \\ | \\ H \end{matrix}-\begin{matrix} H \\ | \\ C \\ | \\ H \end{matrix}-\begin{matrix} X \\ | \\ C^{\bullet} \\ | \\ H \end{matrix}\ \text{etc.}$$

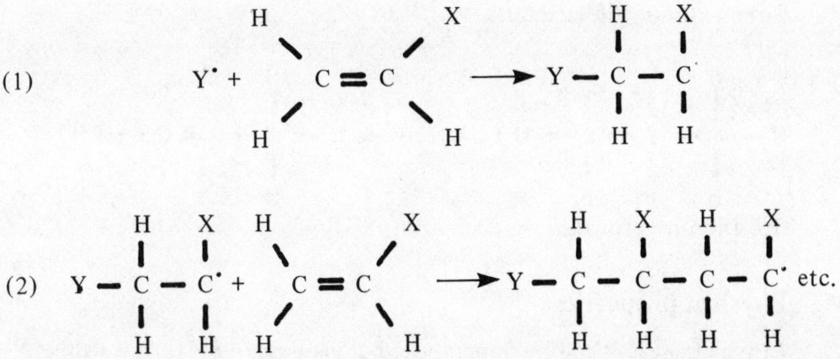

Alkynes

Making alkynes

Ethyne, or acetylene as it is more commonly called, is the most important member of the alkyne series. It is produced industrially by subjecting certain alkanes to very high temperatures. Methane, for example, forms ethyne when heated to about 1500°C.

$$2\ CH_4 \longrightarrow C_2H_2 + 3\ H_2$$

An older method involves firstly the manufacture of calcium carbide. Limestone is decomposed by heat to give quick lime, CaO, which in turn is heated with coke in an electric furnace at over 2000°C. Due to increasing costs of electricity this method is becoming less economical. Ethyne is formed when calcium carbide reacts with water.

$$CaC_2 + 2\ H_2O \longrightarrow C_2H_2 + Ca(OH)_2$$

In general, alkynes can be prepared in the laboratory by warming a mixture of a dibromoalkane with a solution of potassium hydroxide in alcohol. The dibromoalkane should have the bromine atoms on adjacent carbon atoms. For example:

$$\begin{array}{ccccccccccccccc}
 & & H & & H & & H & & & & & & H & & \\
 & & | & & | & & | & & & & & & | & & \\
H & - & C & - & C & - & C & - & H + 2\ OH^- & \longrightarrow & H & - & C & - & C\equiv C - H + 2\ Br^- + 2 \\
 & & | & & | & & | & & & & & & | & & \\
 & & H & & Br & & Br & & & & & & H & &
\end{array}$$

1,2-Dibromopropane Propyne

Physical properties

Ethyne is a gas at room temperature, boiling point −84°C. It is liable to explode if liquefied, but it can be safely stored in cylinders if it is dissolved in propanone (acetone) under pressure.

Chemical properties

Combustion: Ethyne burns in air with a luminous flame producing much soot. This is characteristic of compounds which have such a high proportion of carbon. However, when it is mixed with oxygen in an oxyacetylene burner, much more efficient combustion takes place producing temperatures around 3000°C.

Addition reactions: As with alkenes, alkynes decolourise bromine water and acidified potassium permanganate indicating that they also are **unsaturated.**

Alkynes undergo addition reactions in a two-stage process as shown in the examples given below.

$$\underset{\text{Ethyne}}{H-C\equiv C-H} \xrightarrow{Br_2} \underset{\text{1,2-Dibromoethene}}{(H)(Br)C=C(H)(Br)} \xrightarrow{Br_2} \underset{\text{1,1,2,2-Tetrabromoethane}}{Br-C(H)(Br)-C(H)(Br)-Br}$$

$$\underset{\text{Ethyne}}{H-C\equiv C-H} \xrightarrow{HCl} \underset{\text{Monochloroethene (vinyl chloride)}}{(H)(H)C=C(Cl)(H)} \xrightarrow{HCl} \underset{\text{1,1-Dichloroethane}}{H-C(H)(H)-C(Cl)(H)-Cl}$$

$$\underset{\text{Ethyne}}{H-C\equiv C-H} \xrightarrow[\text{Ni catalyst}]{H_2} \underset{\text{Ethene}}{(H)(H)C=C(H)(H)} \xrightarrow[\text{Ni catalyst}]{H_2} \underset{\text{Ethane}}{H-C(H)(H)-C(H)(H)-H}$$

Like ethene, ethyne can undergo **hydration**, although the conditions are somewhat different. If ethyne is passed into dilute sulphuric acid containing mercury(II) ions and iron(III) ions as catalysts, at about 80°C, it effectively adds on a water molecule to form a compound called ethanal or acetaldehyde. We shall investigate this substance in more detail in chapter 16.

$$H-C\equiv C-H + H_2O \longrightarrow H-C(H)(H)-C(=O)-H$$

Another important reaction of ethyne involves the addition of hydrogen cyanide, HCN, to form acrylonitrile which can be polymerised to polyacrylonitrile, a valuable synthetic fibre.

$$\begin{array}{ccc} H & & CN \\ & \diagdown \quad \diagup & \\ & C = C & \\ & \diagup \quad \diagdown & \\ H & & H \end{array}$$

Acrylonitrile

Aromatic Hydrocarbons

Benzene, C_6H_6, is the simplest member of this group of compounds. All other aromatic compounds can be regarded as being derivatives of it.

Manufacture of benzene

Coal tar is distilled to yield five main fractions. Benzene is present in the lowest boiling fraction which is treated with (1) acid to remove basic compounds such as aniline and (2) alkali to remove acidic compounds such as phenol. It is then dried and redistilled to yield benzene.

The catalytic reforming process for converting aliphatic hydrocarbons obtained from petroleum into aromatic hydrocarbons is becoming increasingly important as a source of benzene. The process involves ring formation and the removal of hydrogen atoms. For example

$$C_6H_{14} \longrightarrow C_6H_6 + 4\,H_2$$

Physical properties

Benzene is a colourless liquid with a smell typical of aromatic compounds. It boils at 80°C and freezes at 5.5°C. It is immiscible with water, but it is miscible with ethanol. It is highly toxic so that great care must be taken when using it.

Chemical properties

Combustion: Benzene burns with a luminous flame producing much soot. This is a characteristic feature of aromatic compounds because of their relatively high proportion of carbon atoms per molecule. If sufficient oxygen is available, benzene will burn completely to form carbon dioxide and water.

Test for unsaturation: Table 14.1 summarises the results of testing various hydrocarbons with bromine water.

Table 14.1

name	formula	effect on bromine water	conclusion
hexane	C_6H_{14}	no reaction	saturated
hexene	C_6H_{12}	decolourised	unsaturated
cyclohexane	C_6H_{12}	no reaction	saturated
cyclohexene	C_6H_{10}	decolourised	unsaturated
benzene	C_6H_6	no reaction	apparently saturated

Since benzene does not decolourise bromine water, its molecule does not contain carbon-carbon double or triple bonds. The structure of benzene was discussed in the previous chapter (page 148).

Substitution: Benzene does not readily react by addition but reacts more easily by substitution. We shall consider only one example here.

If a few drops of liquid bromine are added to benzene in the presence of some iron filings, a reaction occurs in which acidic fumes are produced. The overall effect is the substitution of a hydrogen atom by a bromine atom to yield bromobenzene and hydrogen bromide gas. The iron acts as a catalyst. The mechanism of this reaction is **not** the same as that of the substitution reaction involving alkanes and bromine dealt with earlier in this chapter.

$$\underset{\text{Benzene}}{C_6H_6} + Br_2 \longrightarrow \underset{\text{Bromobenzene}}{C_6H_5Br} + HBr$$

Benzene + Br_2 → Bromobenzene (ring with Br) + HBr

Uses of benzene

1. In the plastics industry: Styrene and cyclohexane are important starting materials for making plastics and are manufactured from benzene. Styrene is made into polystyrene or mixed with another monomer and co-polymerised to produce synthetic rubber. Cyclohexane is required in the manufacture of nylon.

2. Detergents: Benzene is used in the manufacture of detergents of the type known as alkylbenzenesulphonates. The general structure is shown over the page.

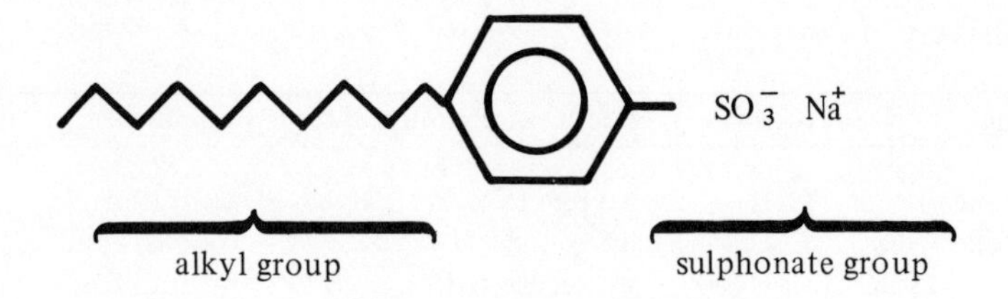

3. Manufacture of insecticides: e.g. DDT, Gammexane (formula – $C_6H_6Cl_6$).

4. Manufacture of important aromatic compounds: e.g. phenol, aniline.

Summary

The reaction of bromine towards the different types of hydrocarbons which we have dealt with in this chapter provides a useful means of comparison between them. Table 14.2 below summarises these reactions.

Table 14.2

	type of reaction	special conditions	example
alkanes	substitution	usually Br_2 vapour; light	$H-CH_2-CH_2-H \xrightarrow{Br_2} H-CH_2-CH_2-Br + HBr$
alkenes	addition	–	$H_2C{=}CH_2 \xrightarrow{Br_2} Br-CH_2-CH_2-Br$
alkynes	addition (two-stage)	–	$H-C{\equiv}C-H \xrightarrow{Br_2} HBrC{=}CBrH \xrightarrow{Br_2} Br-CHBr-CHBr-Br$
aromatic compounds	substitution	liquid Br_2; iron catalyst	$C_6H_6 \xrightarrow{Br_2} C_6H_5Br + HBr$

Examples for practice

1. $C_2H_4 \xrightarrow{Br_2} X \xrightarrow[\text{heat}]{\text{KOH in ethanol}} Y \xrightarrow{HBr} Z$

   X and Z are isomers. Y is unsaturated.

Give the names and structural formulae of X, Y and Z.

2.

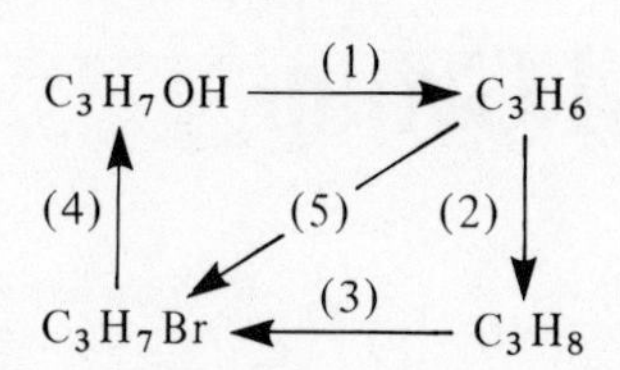

For each of the reactions shown, write down
(a) the type of reaction taking place,
(b) the reagent and conditions, where appropriate, needed to achieve it.

3. C_6H_6 I C_6H_{12} II   Hex-l-ene III   n-Hexane IV

Compounds I and II are both cyclic.
(a) Which pair of the above compounds have the same molecular formula?
(b) Copy down the headings – 'Compound I' and 'Compound II' – and under each heading, write down
  (i) the name of the compound,
  (ii) its effect on bromine water, and
  (iii) those terms which apply to it from the following list
planar ring; puckered ring; aromatic; delocalised electrons.
(c) Give the structural formula of
  (i) an unsaturated isomer of compound III, and
  (ii) an isomer of compound IV.
Beside each formula, give its systematic name.
(d) Which compound – III or IV – will decolourise bromine water? Give the structural formula of the product of this reaction. What other test could be carried out to confirm unsaturation?

$$\text{n-Hexane} \xrightarrow{A} \text{1-bromohexane} \xrightarrow{B} \text{Compound C}$$

(e) What are the conditions required for carrying out reaction A? What type of reaction is this?
(f) Reaction B occurs when 1-bromohexane is warmed in the presence of potassium hydroxide solution.
(i) Name compound C and give its molecular formula.
(ii) In order to show that reaction B has taken place, it is usual practice to add firstly dilute nitric acid and then silver(I) nitrate solution. Explain briefly why both reagents are used.

15 Compounds which Contain the Hydroxyl Group

In this chapter we shall study the chemistry of two types of carbon compounds which contain the hydroxyl group of atoms (–OH), namely
(1) the aliphatic alcohols, and
(2) aromatic hydroxyl compounds, otherwise known as phenols

Alcohols

Table 15.1 Structural formulae of alcohols

name of alcohol	molecular formula	structural formula	type of alcohol
methanol	CH_3OH	$\begin{array}{ccccccc} & & H & & & & \\ & & \mid & & & & \\ H & - & C & - & O & - & H \\ & & \mid & & & & \\ & & H & & & & \end{array}$	primary
ethanol	C_2H_5OH	CH_3CH_2OH $\begin{array}{ccccccccc} & & H & & H & & & & \\ & & \mid & & \mid & & & & \\ H & - & C & - & C & - & O & - & H \\ & & \mid & & \mid & & & & \\ & & H & & H & & & & \end{array}$	primary
propan-1-ol	C_3H_7OH	$CH_3CH_2CH_2OH$ $\begin{array}{ccccccccccc} & & H & & H & & H & & & & \\ & & \mid & & \mid & & \mid & & & & \\ H & - & C & - & C & - & C & - & O & - & H \\ & & \mid & & \mid & & \mid & & & & \\ & & H & & H & & H & & & & \end{array}$	primary

name of alcohol	molecular formula	structural formula	type of alcohol
propan-2-ol	C_3H_7OH	$\begin{array}{c} CH_3\underset{\mid}{C}HCH_3 \\ OH \end{array}$ $\begin{array}{ccccccc} & & H & & H & & H & \\ & & \mid & & \mid & & \mid & \\ H & - & C & - & C & - & C & - & H \\ & & \mid & & \mid & & \mid & \\ & & H & & OH & & H & \end{array}$	secondary
butan-1-ol	C_4H_9OH	$CH_3CH_2CH_2CH_2OH$	primary
butan-2-ol	C_4H_9OH	$\begin{array}{c} CH_3CH_2\underset{\mid}{C}HCH_3 \\ OH \end{array}$	secondary
2-methyl propan-2-ol	C_4H_9OH	$\begin{array}{ccccc} & & CH_3 & & \\ & & \mid & & \\ CH_3 & - & C & - & CH_3 \\ & & \mid & & \\ & & OH & & \end{array}$	tertiary

In a primary alcohol, the hydroxyl group is attached to the end of a chain of carbon atoms; whereas in a secondary alcohol, it is attached to an intermediate carbon atom. In a tertiary alcohol the $-OH$ group is attached to an intermediate carbon atom, which *also* has a hydrocarbon branch attached. Table 15.2 summarises these differences.

Table 15.2

	primary	secondary	tertiary
characteristic grouping of atoms	$-CH_2-OH$	$\begin{array}{l} \diagdown \\ \quad CH-OH \\ \diagup \end{array}$	$\begin{array}{ccc} & \mid & \\ - & C & - OH \\ & \mid & \end{array}$
general formula of alcohol	$R-CH_2-OH$	$\begin{array}{l} R \\ \quad \diagdown \\ \qquad CH-OH \\ \quad \diagup \\ R' \end{array}$	$\begin{array}{ccccc} & & R' & & \\ & & \mid & & \\ R & - & C & - & OH \\ & & \mid & & \\ & & R'' & & \end{array}$

R, R′ and R″ are variable alkyl groups which can be the same or different. **Note**: the *number* of H atoms bonded to the C atom which 'carries' the –OH group.

Isomerism

From table 15.1 we can see that there are two possible structures for propanol. The difference lies in the position of the –OH group. When naming an alcohol, therefore, it is important to specify the carbon atom to which the –OH group is attached. The name of the last compound listed on table 15.1 is related to its structure as follows:

2 – Methyl	propan	2 – ol
methyl branch attached to 2nd C atom in chain	chain consists of 3 C atoms	–OH group joined to 2nd C atom in the chain

$$\begin{array}{c} CH_3 \\ | \\ CH_3 - C - CH_3 \\ | \\ OH \end{array}$$

Number of –OH groups per molecule

All of the alcohols mentioned so far possess one hydroxyl group per molecule, and, as such, are referred to as **monohydric** alcohols. **Polyhydric** alcohols (i.e. those containing more than one –OH group per molecule) do exist, however, and two important examples are given below.

Table 15.3

	a dihydric alcohol	a trihydric alcohol
formula	$(CH_2OH)_2$; $\begin{array}{cc} CH_2 - & CH_2 \\ \mid & \mid \\ OH & OH \end{array}$	$C_3H_5(OH)_3$; $\begin{array}{ccc} CH_2 - & CH - & CH_2 \\ \mid & \mid & \mid \\ OH & OH & OH \end{array}$
common name	ethylene glycol	glycerol or glycerine
systematic name	ethan – 1,2 – diol	propan – 1,2,3 – triol
importance	it is used as an anti-freeze and in manufacture of polyester resins and 'terylene'.	it is produced when fats and oils are hydrolysed in soap manufacture. Used in making explosives.

Manufacture of alcohols

Ethanol is produced from carbohydrates in brewing. This is summarised below.

$$\underset{\text{(polysaccharide)}}{\text{Starch}} \xrightarrow{\text{hydrolysis}} \underset{\substack{C_{12}H_{22}O_{11}\\ \text{(disaccharide)}}}{\text{Maltose}} \xrightarrow{\text{hydrolysis}} \underset{\substack{C_6H_{12}O_6\\ \text{(monosaccharide)}}}{\text{Glucose}} \xrightarrow{\text{fermentation}} \underset{C_2H_5OH}{\text{Ethanol}} + CO_2$$

The various steps are catalysed by enzymes. The resulting solution contains about 10% ethanol and this can be concentrated by fractional distillation.

As mentioned earlier (page 158), ethene can be converted to ethanol by hydration and other alcohols can be prepared by this type of reaction.

$$H_2C{=}CH_2 \xrightarrow[\text{conc. } H_2SO_4]{\text{addition of}} H{-}CH_2{-}CH_2{-}O{-}SO_3H \xrightarrow[\text{with water}]{\text{hydrolysis}} H{-}CH_2{-}CH_2{-}OH$$

Properties of alcohols

The hydroxyl group has a considerable influence on the physical and chemical properties of alcohols, especially the earlier members of the series.

1. **Boiling point.** As can be seen from Fig. 15.1 the boiling points of

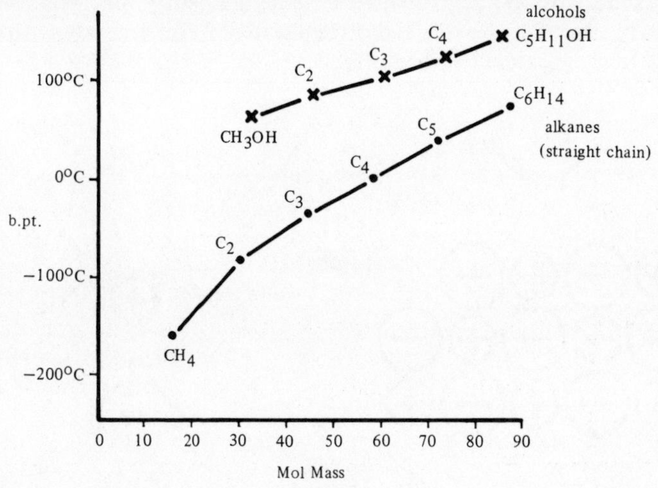

Fig. 15.1: boiling point v molecular mass

alcohols are relatively high compared with other carbon compounds of similar molecular weight. Hydrogen bonding between molecules is responsible for this effect (Fig. 15.2).

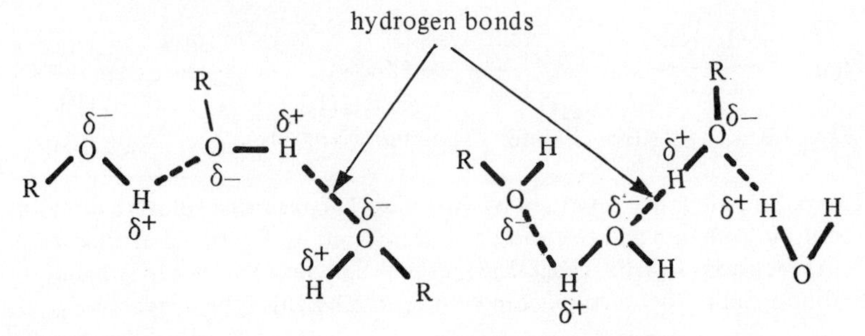

Fig. 15.2 hydrogen bonding between molecules

Fig. 15.3 miscibility with water

2. Miscibility with water. Hydrogen bonding between alcohol and water molecules (Fig. 15.3) enables the first few members of the alcohol series to be miscible with water. This is in contrast to the majority of carbon compounds.

3. Acidity. Water ionises only to a very slight extent. Ethanol has an even lesser tendency to dissociate into ions and is neutral to litmus.

4. Action as a polar solvent. The polarity of the water molecule plays a vital role in its ability to break down the crystal lattice of an ionic compound and so dissolve it. Alcohols also show this ability, albeit to a lesser degree. Potassium hydroxide and silver nitrate, for example, are both soluble to some extent in ethanol.

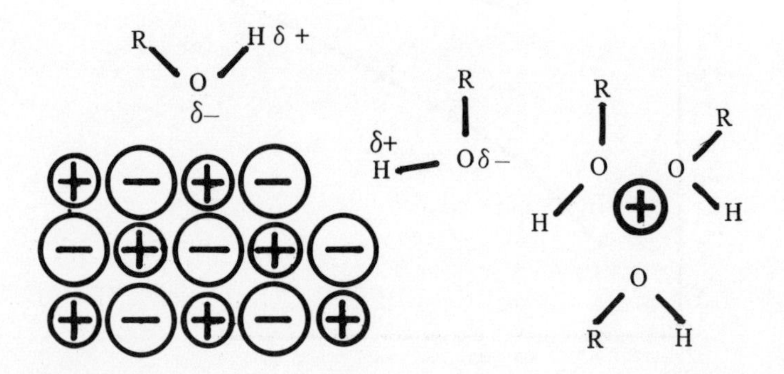

Fig. 15.4

5. Reaction with sodium metal. Sodium reacts vigorously with water to release hydrogen gas and form an alkaline solution of sodium hydroxide.

$$2Na + 2H_2O \longrightarrow 2Na^+_{(aq)} + 2OH^-_{(aq)} + H_{2(g)}$$

Ethanol has a similar reaction with sodium, though less vigorous. Hydrogen is released and a substance called **sodium ethoxide** is formed, which, like sodium hydroxide, is a strong base and is alkaline to litmus.

$$2Na + 2C_2H_5OH \longrightarrow \underbrace{2Na^+ + 2C_2H_5O^-}_{\text{sodium ethoxide}} + H_{2(g)}$$

Other alcohols show similar behaviour.
Note: only the hydrogen atom of the –OH group is displaced in this reaction.

6. Reaction with phosphorus pentachloride (PCl_5). This substance is a white solid which reacts violently with water to give off steamy fumes of hydrogen chloride.

$$PCl_5 + H_2O \longrightarrow 2HCl_{(g)} + POCl_3$$

A similar reaction occurs with alcohols. Ethanol reacts vigorously with PCl_5 to form chloroethane and fumes of hydrogen chloride.

$$PCl_5 + C_2H_5OH \longrightarrow \underbrace{C_2H_5Cl}_{\text{chloroethane}} + HCl_{(g)} + POCl_3$$

In this reaction, the –OH group is entirely removed and replaced by a chlorine atom.

The hydroxyl group is chiefly responsible for the properties and reactions which we have considered above, so that the strong resemblance between water and alcohols is only to be expected.

We shall deal with two other important reactions of alcohols. These are:
(1) the formation of esters, which we will consider in the next section of this chapter and
(2) the oxidation of alcohols, which we will leave until chapter 16.

Esters

Esterification – the formation of esters

A mixture of ethanol and concentrated ethanoic acid (commonly known as glacial acetic acid) is warmed gently for a few minutes in the presence of one or two drops of concentrated sulphuric acid. When this mixture is poured onto some cold water in a beaker, an oily film with a pleasant fruity smell is observed on the surface of the water. This substance is called **ethyl ethanoate** (ethyl acetate) and is an example of an **ester.** It is formed by a condensation reaction between ethanol and ethanoic acid. The reaction is reversible.

$$\underset{\text{ethanoic acid}}{CH_3COOH} + \underset{\text{ethanol}}{C_2H_5OH} \rightleftharpoons \underset{\text{ethyl ethanoate}}{CH_3COOC_2H_5} + H_2O$$

$$CH_3-C\begin{matrix} \nearrow O \\ \searrow \boxed{O-H \quad H}-O \end{matrix} \diagup C_2H_5 \rightleftharpoons CH_3-C\begin{matrix} \nearrow O \\ \searrow O-C_2H_5 \end{matrix} + H_2O$$

The sulphuric acid performs two important functions in this reaction.
(1) It provides hydrogen ions, which catalyse the reaction.
(2) By removing water, it shifts the equilibrium position to the right to give an improved yield of ester.

By varying the alcohol and/or the carboxylic acid, other esters can be prepared using the method outlined above.

Table 15.4

carboxylic acid	alcohol	ester produced
propanoic, C_2H_5COOH	methanol, CH_3OH	methyl propanoate, $C_2H_5COOCH_3$
ethanoic, CH_3COOH	pentanol, $C_5H_{11}OH$	pentyl ethanoate, $CH_3COOC_5H_{11}$
benzoic, C_6H_5-COOH	ethanol, C_2H_5OH	ethyl benzoate, $C_6H_5-COOC_2H_5$

carboxylic acid	alcohol	ester produced
butanoic, C_3H_7COOH	methanol, CH_3OH	methyl butanoate, $C_3H_7COOCH_3$
salicylic	methanol	methyl salicylate ('oil of wintergreen')

The formation of esters can be summarised as follows:

$$\begin{array}{ccccc} R-COOH & + R'-OH & \rightleftharpoons & R-COO-R' & + H_2O \\ \text{acid} & + \text{alcohol} & \rightleftharpoons & \text{ester} & + \text{water} \end{array}$$

In esterification, the oxygen for the water comes from the acid. This has been shown by reacting ethanol containing the heavier isotope, ^{18}O, with ethanoic acid. Analysis of the products, making use of a mass spectrometer, has shown that the water formed contains only ^{16}O.

$$CH_3-C(=O)-OH + H-{}^{18}O-C_2H_5 \longrightarrow CH_3-C(=O)-{}^{18}O-C_2H_5 + H_2O$$

Uses of esters

1. **Solvents**: ethyl ethanoate is used as a solvent for adhesives and nail varnish.
2. **Flavouring**: e.g. 'pear drops' (pentyl ethanoate), banana (methyl butanoate).
3. **Perfumes**: e.g. ethyl benzoate.
4. **Medications**: e.g. methyl salicylate.

The hydrolysis of esters

Esters can be broken down or hydrolysed by heating them in the presence of strong acids or alkalis. Acid hydrolysis is reversible and is essentially the reverse of esterification.

$$\underset{\text{ester}}{R-C(=O)-O-R'} + H_2O \xrightarrow{H^+} \underset{\text{acid}}{R-C(=O)-O-H} + \underset{\text{alcohol}}{R'-OH}$$

Alkaline hydrolysis is not reversible since the salt of the acid is formed. Hence, this is a more efficient method.

$$\underset{\text{ester}}{R-C\overset{\displaystyle O}{\underset{\displaystyle O-R'}{\lessgtr}}} + Na^{+} + OH^{-} \longrightarrow \underset{\text{sodium salt of acid}}{R-COO^{-} + Na^{+}} + \underset{\text{alcohol}}{R'-OH}$$

The hydrolysis of an ester is best achieved by heating the ester with an alkali for several minutes under reflux. Figs. 15.5 and 15.6 illustrate two

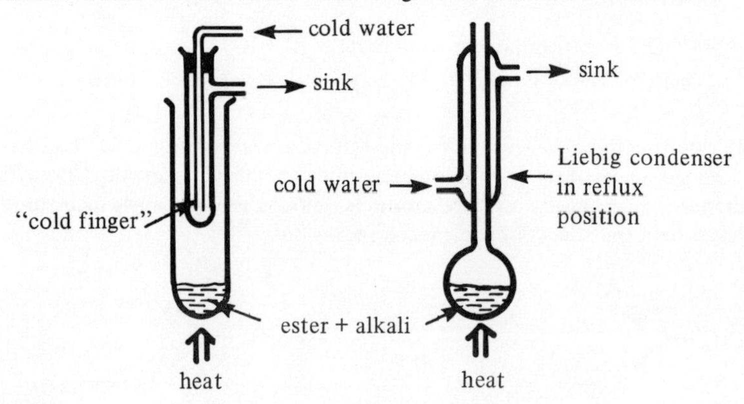

Fig. 15.5

Fig. 15.6

possible methods of doing this. After about half an hour, most of the ester will be hydrolysed to the salt and alcohol. For example:

$$\underset{\text{ethyl ethanoate}}{CH_3COOC_2H_{5(l)}} + Na^{+}_{(aq)} + OH^{-}_{(aq)} \longrightarrow \underset{\text{sodium ethanoate}}{CH_3COO^{-}_{(aq)} + Na^{+}_{(aq)}} + \underset{\text{ethanol}}{C_2H_5OH_{(}}$$

The alcohol can then be removed by distillation. The salt of the acid remains in solution and can be converted to the acid by adding a strong, inorganic acid. For example:

$$CH_3COO^{-}_{(aq)} + H^{+}_{(aq)} \longrightarrow CH_3COOH_{(l)}$$

The alkaline hydrolysis of esters is called **saponification**, as it is the same process as that involved in the manufacture of soap. Fats and oils are esters of long chain fatty acids (such as stearic acid, $C_{17}H_{35}COOH$, and oleic acid, $C_{17}H_{33}COOH$) and the polyhydric alcohol, glycerol. When these esters, known as glycerides, are heated with alkali, hydrolysis occurs forming glycerol and soap, which consists of the salts of the long chain fatty acids.

Fat or oil + alkali $\longrightarrow$ Glycerol + soap
e.g. Na^+OH^- e.g. $C_{17}H_{35}COO^-Na^+$
(sodium stearate)

Phenols

Aromatic compounds which contain one or more hydroxyl groups attached to a benzene ring are classified as phenols. The simplest member is itself called phenol, molecular formula – C_6H_5OH, structural formula:

$$C_6H_5-\overset{\delta-}{O}-\overset{\delta+}{H}$$

It is a solid at room temperature (m. pt. 42°C) and it is slightly soluble in water. It is highly caustic and contact with the skin should be avoided as it produces blisters.

The aqueous solution of phenol is weakly acidic to pH paper. Phenol used to be known as carbolic acid. The solution reacts very slowly with a fresh piece of magnesium ribbon. It does not displace carbon dioxide from a carbonate, since it is a weaker acid than carbonic acid.

The effect of water on phenol can be expressed as follows:

$$C_6H_5-O-H \rightleftharpoons C_6H_5-O^-_{(aq)} + H^+_{(aq)}$$

In water, the equilibrium lies well over to the left, i.e. phenol dissociates into ions to a very slight extent.

Phenol is much more soluble in alkalis, since this causes the equilibrium to shift to the right. The characteristic smell of phenol disappears on dissolving it in alkali. For example:

$$C_6H_5-O-H + Na^+_{(aq)} + OH^-_{(aq)} \longrightarrow \underbrace{C_6H_5-O^-_{(aq)} + Na^+_{(aq)}}_{\text{sodium phenoxide or sodium phenate}} + H_2O$$

If acid is added to this solution, the equilibrium position is shifted back to the left and phenol is reformed. Evidence for this is the formation of oily droplets and the return of the characteristic smell.

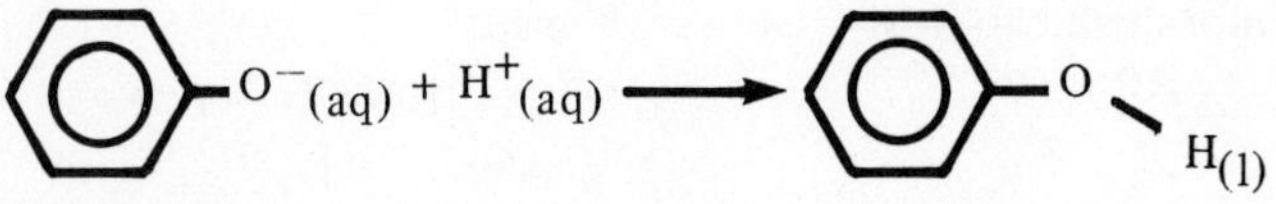

$$C_6H_5O^-_{(aq)} + H^+_{(aq)} \longrightarrow C_6H_5OH_{(l)}$$

Uses of phenols

Phenol is the starting material for a number of very useful substances. The main use is in the manufacture of Bakelite, a thermosetting plastic made by condensing phenol with formaldehyde. Phenol is also converted by a series of reactions into the monomers required in the manufacture of nylon.

Phenol is a powerful germicide. This discovery was made in 1865 by Joseph Lister but since then, it has been found that compounds derived from phenol are more effective and safer to use, e.g. 'dettol' and 'TCP' – *t*ri*c*hloro*p*henol – formula:

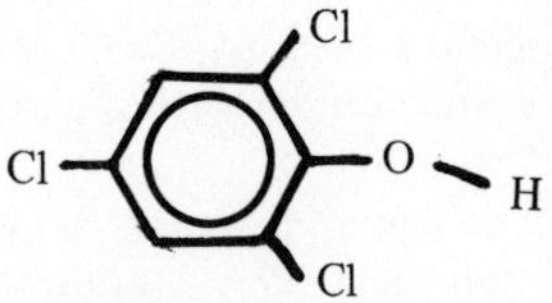

(2,4,6-trichlorophenol: benzene ring with Cl, Cl, Cl and O–H)

Other uses of phenol include the manufacture of aspirin, dyes, wood preservatives and selective weedkillers.

Examples for practice

1. There is one other isomer of butanol not listed in table 15.1. Draw its structural formula. Deduce its name and decide what type of alcohol it is.

2. Give the structural formula and type of each of the following alcohols. (a) Pentan-3-ol, (b) 2-methylbutan-1-ol, (c) 3-ethylpentan-3-ol.

3. Give the name and type of the following alcohols.

(a) $CH_3CH_2 - C(CH_3)(OH) - CH_3$

(b) $CH_3CH_2 - C(CH_3)(CH_3) - CH_2OH$

(c) $CH_3CH(CH_3)CH(OH)CH_3$

4. A compound, X, has the following percentage composition by mass: 60% C; 13.3% H; 26.7% O.
(a) Calculate its empirical formula.
(b) Molecular mass of X is 60. What is its molecular formula?
(c) X reacts with (i) sodium metal and (ii) phosphorus pentachloride to release different gases. What are these gases? What information do these reactions give us about X?
(d) From your answers to (a), (b) and (c) above, draw the two possible structural formulae for X and write down their systematic names.
(e) There is a third isomer of X which does *not* react with either sodium metal or phosphorus pentachloride. What does this tell us about this isomer?

5. (a) Copy and complete the following table.

	alcohol		acid		compound formed
(1)	methanol	+	ethanoic acid	⟶	————————
(2)	——————	+	—————————	⟶	propyl benzoate
(3)	butanol	+	—————————	⟶	———— methanoate
(4)	——————	+	—————————	⟶	ethyl propanoate

(b) What *type* of compound is formed in the above reaction? What other substance is also formed? Concentrated sulphuric acid is often used in this reaction. Give one reason for its use.
(c) Choose one of the reactions from the list in (a) and write the structural formulae of the alcohol, acid and compound formed. Show clearly how the alcohol and acid combine.
(d) The above type of reaction can be reversed. What name is given to this reverse process? Name a chemical that can be used to achieve this. Why is this process of particular importance when applied to fats and oils?
(e) When 0.1 mole of a certain alcohol was burnt completely 8.96 litres of carbon dioxide (measured at STP) was produced. Name the alcohol and write the equation for its complete combustion. How many moles of oxygen would have been required for the above reaction?

16 The Oxidation of Alcohols

At an earlier stage of your study of Chemistry you will have discovered that ethanol can be oxidised to give ethanoic acid (acetic acid). This can be achieved using a variety of chemical reagents or by certain bacteria. Vinegar can be made by the bacterial oxidation of the alcohol present in poor quality wines.

In this chapter we shall be concerned with a more thorough investigation of the oxidation of alcohols by chemical means. The combustion of alcohols to form carbon dioxide and water is essentially an oxidation process and we have referred earlier (page 31) to the relationship between the heat of combustion of alcohols and their molecular structure. However, we shall be considering here less extensive oxidations which result in forming new carbon compounds other than oxides of carbon.

Method of Oxidation

Several oxidising agents can be used to oxidise alcohols including powerful reagents such as potassium permanganate and potassium dichromate acidified with dilute sulphuric acid. Oxidation can also be achieved catalytically when an alcohol-air mixture is passed over heated copper.

A convenient method to use in the laboratory involves passing alcohol vapour over hot copper(II) oxide as shown in Fig. 16.1 below. In this experiment, the oxidation of an alcohol is accompanied by reduction of the copper(II) oxide to copper metal.

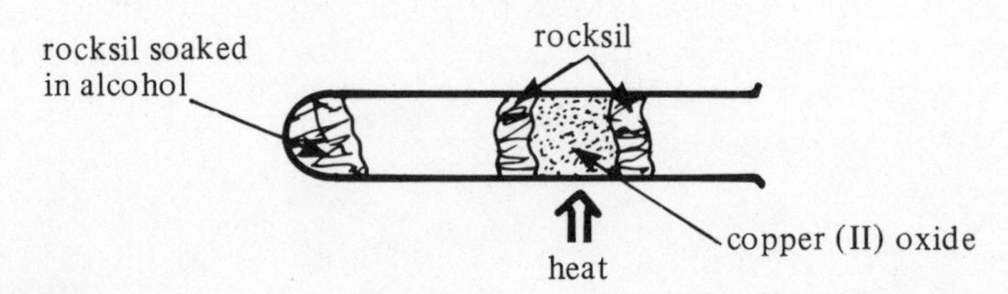

Fig. 16.1

Products of Oxidation

When primary and secondary alcohols are tested in this way, the copper (II) oxide is clearly reduced to copper. Tertiary alcohols are difficult to oxidise and do not give this reaction.

Primary alcohols

When ethanol is oxidised by this method a very sharp-smelling vapour is produced. This substance is known as **ethanal** (acetaldehyde). Other primary alcohols yield similar compounds which are known generally as **aldehydes.**

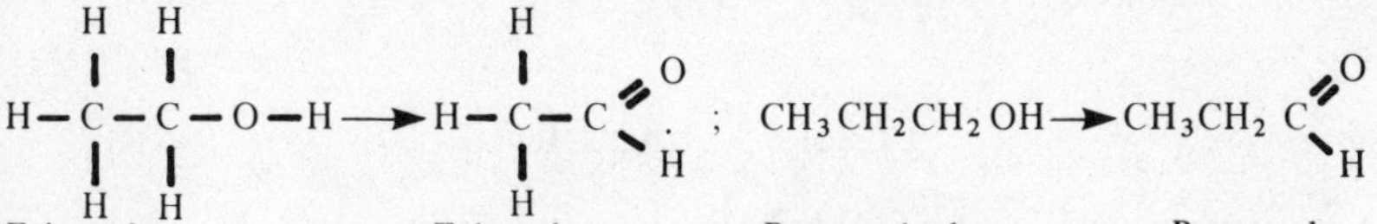

$$\text{H}-\underset{\text{H}}{\overset{\text{H}}{\text{C}}}-\underset{\text{H}}{\overset{\text{H}}{\text{C}}}-\text{O}-\text{H} \longrightarrow \text{H}-\underset{\text{H}}{\overset{\text{H}}{\text{C}}}-\text{C}{\overset{\text{O}}{\underset{\text{H}}{\lessgtr}}} \;;\quad CH_3CH_2CH_2OH \longrightarrow CH_3CH_2C{\overset{\text{O}}{\underset{\text{H}}{\lessgtr}}}$$

Ethanol Ethanal Propan-1-ol Propanal

In general,

$$\text{Primary alcohols}\quad \text{R}-\underset{\text{H}}{\overset{\text{H}}{\text{C}}}-\text{O}-\text{H} \longrightarrow \text{R}-\text{C}{\overset{\text{O}}{\underset{\text{H}}{\lessgtr}}}\quad \text{Aldehydes}$$

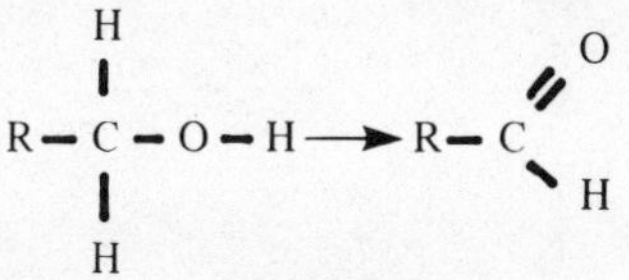

Overall equation for the reaction:

Reduction (CuO → Cu)

$$CuO + RCH_2OH \longrightarrow Cu + RCHO + H_2O$$

Oxidation (RCH_2OH → RCHO)

Secondary alcohols

When a secondary alcohol is oxidised by the above method a somewhat more pleasant and less pungent vapour is formed. Secondary alcohols yield compounds known generally as **ketones.** The simplest ketone is **propanone** (acetone) obtained on the oxidation of propan-2-ol, as shown over the page.

$$\mathrm{H{-}\overset{H}{\underset{H}{C}}{-}\overset{H}{\underset{OH}{C}}{-}\overset{H}{\underset{H}{C}}{-}H \longrightarrow H{-}\overset{H}{\underset{H}{C}}{-}\underset{\|\,O}{C}{-}\overset{H}{\underset{H}{C}}{-}H}$$

Propan-2-ol → Propanone

or

$$\mathrm{(CH_3)_2CHOH \longrightarrow (CH_3)_2C{=}O}$$

Propan-2-ol → Propanone

$$\mathrm{CH_3CH_2\underset{\displaystyle |\atop OH}{CH}CH_3 \longrightarrow CH_3CH_2\underset{\displaystyle \|\atop O}{C}CH_3}$$

Butan-2-ol → Butan-2-one

In general,

Secondary alcohols $\mathrm{RR'CHOH \longrightarrow RR'C{=}O}$ Ketones

Overall equation for the reaction:

$$\mathrm{CuO + RR'CHOH \longrightarrow Cu + RR'C{=}O + H_2O}$$

(Reduction: CuO → Cu; Oxidation: RR'CHOH → RR'C=O)

Further Oxidation

To find out whether or not further oxidation is possible, the same method is used except that this time the alcohol is replaced by an aldehyde or ketone.

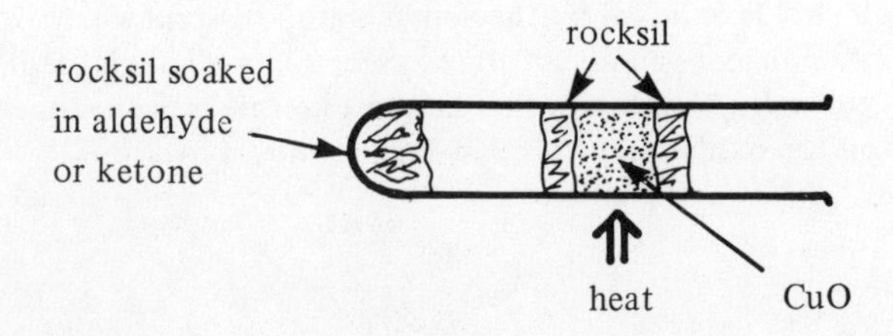

Fig. 16.2

Aldehydes

When ethanal is tested, the copper(II) oxide is reduced and an acidic vapour with a smell of vinegar is produced. The product is, in fact, ethanoic acid (acetic acid). Other aldehydes yield similar products.

$$\underset{\text{Ethanal}}{CH_3CHO} \longrightarrow \underset{\text{Ethanoic acid}}{CH_3COOH}\;;\quad \underset{\text{Butanal}}{CH_3CH_2CH_2CHO} \longrightarrow \underset{\text{Butanoic Acid}}{CH_3CH_2CH_2COOH}$$

In general,

$$\text{Aldehydes}\quad R{-}C(=O){-}H \longrightarrow R{-}C(=O){-}O{-}H \quad \text{Carboxylic Acids}$$

Overall equation for the reaction:

$$CuO + R{-}CHO \longrightarrow Cu + R{-}COOH$$

(Reduction: CuO → Cu; Oxidation: R—CHO → R—COOH)

Ketones

If a ketone, e.g. propanone, is tested, no apparent reduction of the copper(II) oxide occurs indicating that ketones are not readily oxidised. If ketones (and also tertiary alcohols) are to be oxidised, carbon-to-carbon bond breakage must first take place. Prolonged heating with powerful oxidising agents such as acidified permanganate or dichromate will achieve this, but such reactions are too complex to consider at this stage.

Summary

1. Primary alcohols can undergo **two** stages of oxidation as follows.

$$\underset{R{-}CH_2OH}{\text{Primary Alcohols}} \longrightarrow \underset{R{-}CHO}{\text{Aldehydes}} \longrightarrow \underset{R{-}COOH}{\text{Carboxylic Acids}}$$

2. Secondary alcohols can readily undergo only **one** stage of oxidation, namely –

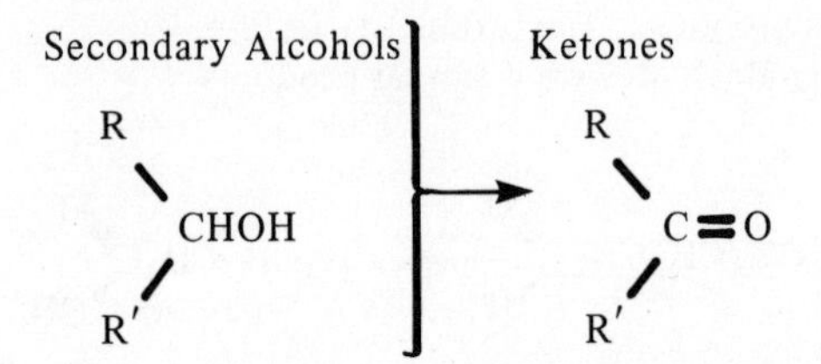

3. Tertiary alcohols are not readily oxidised.

Note: (a) The following points justify the above changes as being oxidation processes.

(i) In bringing about these changes, we observe the reduction of copper(II) oxide.

(ii) We can make use of an older definition of oxidation, namely, that it involves an increase in the proportion of oxygen in a compound or a decrease in the proportion of hydrogen. Thus:

$$R-\underset{H}{\overset{H}{\underset{|}{\overset{|}{C}}}}-O-H \xrightarrow[\text{removed}]{\text{Two H atoms}} R-C\begin{matrix} /\!\!/ O \\ \backslash H \end{matrix} \xrightarrow[\text{added}]{\text{One O atom}} R-C\begin{matrix} /\!\!/ O \\ \backslash O-H \end{matrix}$$

(b) The difference between aldehydes and ketones.

Both types of compound possess a carbon-to-oxygen double bond (C = O), which is known as the **carbonyl** group. An aldehyde has the carbonyl group at the end of the carbon chain and has a hydrogen atom attached to it.

In a ketone the carbonyl group is joined to carbon atoms on both sides. This difference in chemical structure gives rise to some differences in properties which we shall investigate in the next section.

Chemical Tests to Distinguish Aldehydes and Ketones

We have already seen that aldehydes can reduce copper(II) oxide, while ketones cannot. Other tests can be carried out to distinguish aldehydes and ketones to illustrate further this difference in reducing power.

Table 16.1

reagent	observation	explanation including the ion-electron equation for the oxidising agent
(1) Fehling's or Benedict's solution	aldehyde: blue solution $\longrightarrow$ red precipitate ketone: no effect	aldehyde reduces Cu^{2+} ions in alkaline solution to copper(I) oxide, Cu_2O. $Cu^{2+} + e \longrightarrow Cu^{+}$ $[\longrightarrow Cu_2O]$
(2) Tollen's reagent (ammoniacal silver(I) nitrate)	aldehyde: silver mirror formed ketone: no effect	aldehyde reduces Ag^{+} ions to silver atoms forming thin layer inside test-tube. $Ag^{+} + e \longrightarrow Ag$
(3) acidified potassium dichromate	aldehyde: orange solution turns blue-green ketones: no effect	aldehyde reduces dichromate ions (orange) to chromium(III) ions (blue-green) $Cr_2O_7^{2-} + 14H^{+} + 6e \longrightarrow 2\,Cr^{3+} + 7\,H_2O$

These tests show that aldehydes act as reducing agents or electron donors to a variety of reagents, whereas ketones do not. Oxidation occurs readily where a hydrogen atom is attached directly to a carbonyl group as in an aldehyde.

In test (3), which is carried out in acidic conditions, an aldehyde is oxidised to the corresponding carboxylic acid. For example:

$$\underset{\text{ethanal}}{CH_3CHO} + H_2O \longrightarrow \underset{\text{ethanoic acid}}{CH_3COOH} + 2\,H^{+} + 2e \text{ (to the dichromate ions)}$$

In tests (1) and (2), which are carried out in alkaline conditions, an aldehyde is oxidised to the anion of the corresponding carboxylic acid. For example:

$$\underset{\text{ethanal}}{CH_3CHO} + 3\,OH^{-} \longrightarrow \underset{\text{ethanoate ion}}{CH_3COO^{-}} + 2\,H_2O + 2e \text{ (to } Cu^{2+} \text{ or } Ag^{+} \text{ ions)}$$

Our previous studies of carbohydrates showed that glucose can reduce Fehling's or Benedict's solutions, whereas sucrose cannot. However, if sucrose is hydrolysed and then tested, reduction of these reagents can occur.

The monosaccharide glucose has a ring structure which can be easily 'opened' to give a chain structure. The open chain structure of glucose has an aldehyde group of atoms

$\left(-C \overset{\displaystyle O}{\underset{\displaystyle H}{\lessgtr}}\right)$ and this gives glucose its reducing properties.

Sucrose is a disaccharide in which the two monosaccharide units are linked together in such a way that there is no 'free' aldehyde group and hence the molecule does not have reducing properties. Hydrolysis of sucrose yields glucose and fructose, both of which are reducing sugars.

Maltose is both a disaccharide and a reducing agent since its units are linked so as to leave a 'free' aldehyde group.

Carboxylic Acids

We have already noted that when aldehydes are oxidised they yield compounds known as carboxylic acids. These compounds contain the **carboxyl** group of atoms,

i.e. $-C \overset{\displaystyle O}{\underset{\displaystyle O-H}{\lessgtr}}$. The first three members of the series are:-

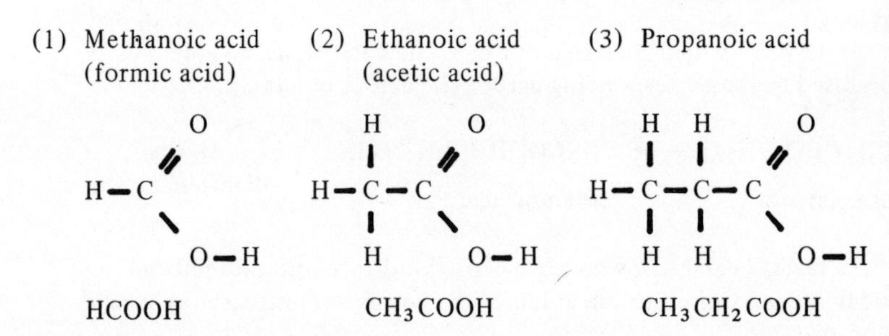

1. Physical Properties

As with alcohols, hydrogen bonding can occur between molecules so that these compounds have relatively high boiling points and melting points. The first five or six members are also miscible with water.

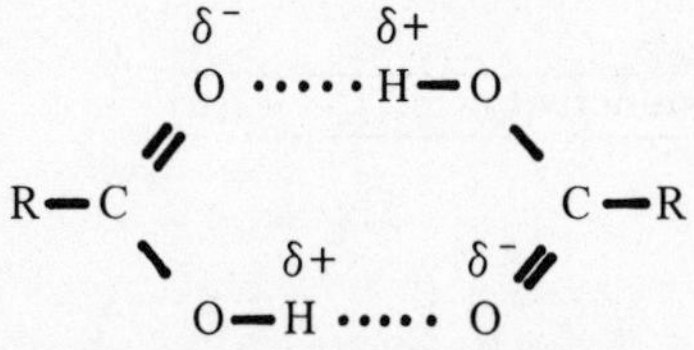

$$
\begin{array}{ccccccc}
 & & \delta^- & & \delta+ & & \\
 & & O & \cdots\cdots & H-O & & \\
 & /\!/ & & & & \backslash & \\
R-C & & & & & & C-R \\
 & \backslash & \delta+ & & \delta^- & /\!/ & \\
 & & O-H & \cdots\cdots & O & &
\end{array}
$$

••••• = hydrogen bonds

Hydrogen bonding can result in a sort of 'double link' between the acid molecules as shown above. Such pairs of molecules are sometimes called **dimers**.

2. Chemical properties in aqueous solution

Carboxylic acids have a much greater tendency than alcohols to release hydrogen ions when added to water. The C=O part of the molecule exerts a strong electron pull thus weakening the O – H bond. Furthermore, after the proton has been removed, the extra electron is 'shared' by the remaining atoms of the group as shown below.

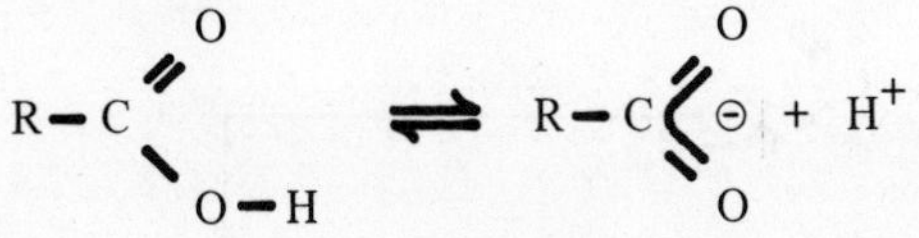

$$R-C\begin{matrix} /\!/ O \\ \backslash O-H \end{matrix} \rightleftharpoons R-C\begin{matrix} O \\ \ominus \\ O \end{matrix} + H^+$$

An aqueous solution of ethanoic acid gives typical reactions of dilute acids,

(1) neutralising alkalis to form salts, e.g.

$$CH_3COOH + NaOH \longrightarrow \underset{\text{sodium ethanoate}}{CH_3COONa} + H_2O$$

(2) releasing carbon dioxide from carbonates, e.g.

$$2CH_3COOH + CuCO_3 \longrightarrow (CH_3COO)_2Cu + CO_2 + H_2O$$

(3) releasing hydrogen gas with reactive metals, e.g.

$$2CH_3COOH + Mg \longrightarrow (CH_3COO)_2Mg + H_2$$

Compared with the mineral acids, however, carboxylic acids are very much weaker. Experimental evidence for this can be obtained by comparing conductivities and pH values of solutions of equal concentration. Typical results are shown in table 16.2.

Table 16.2

solution	conductivity (mA)	pH
0.1M HCl	100	1
0.1M CH_3COOH	30	4

Hydrochloric acid is a strong acid since it is fully dissociated into ions. However, ethanoic acid is a weak acid since it is only partially dissociated into ions and there are many molecules of the acid still left in solution.

$$CH_3COOH \rightleftharpoons CH_3COO^- + H^+.$$

3. Chemical properties of the concentrated acids

Previously you will have noted the marked difference in chemical properties of other acids e.g. H_2SO_4, HNO_3 when dilute and when concentrated. Pure carboxylic acids also have different properties from their aqueous solutions. Two of these reactions are illustrated below. Note that in both of these reactions it is the C – O bond which is broken and that the –OH group is removed.

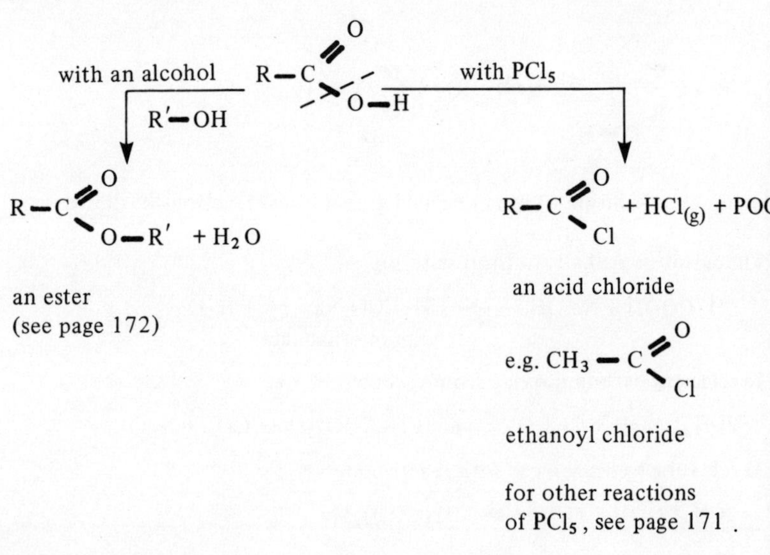

Examples for practice. These are to be found at the end of Chapter 17.

17 Amines

Ammonia, NH_3, is a very important chemical which you will have investigated in some detail earlier in the course. If one or more of the hydrogen atoms in the ammonia molecule are replaced by a hydrocarbon group, the new compound is known as an **amine**.

Structure and Formulae of Amines

1. Ammonia

The shape of an ammonia molecule is shown below.

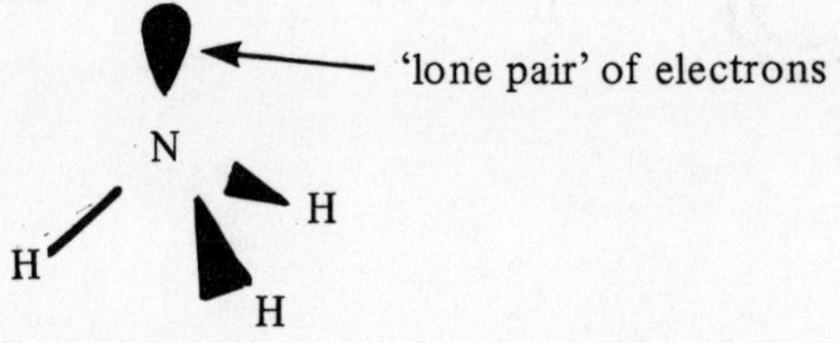

The nitrogen atom has a 'lone pair' of electrons, i.e. a full electron-pair cloud which is not involved in bonding with other atoms. This 'lone pair' repels the bonded pairs of electrons giving ammonia a pyramidal shape as shown above. When a hydrogen atom is replaced by a hydrocarbon group, the shape of the molecule in the vicinity of the nitrogen atom is not altered to any great extent.

2. Aliphatic amines

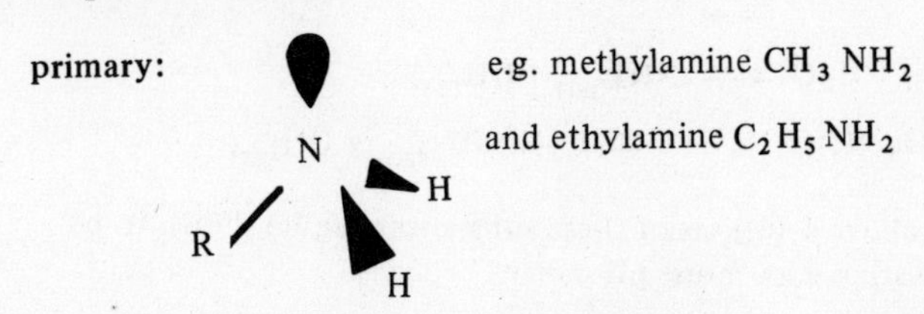

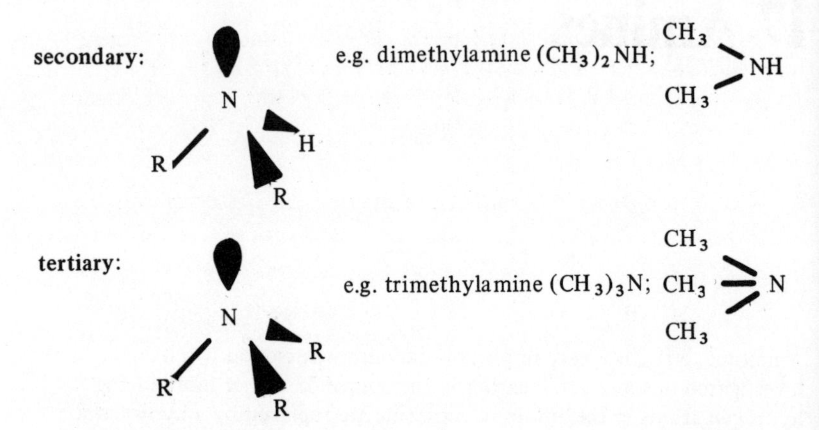

3. Aromatic amines

Aniline or phenylamine is the simplest aromatic amine. It is a primary amine, molecular formula – $C_6H_5NH_2$

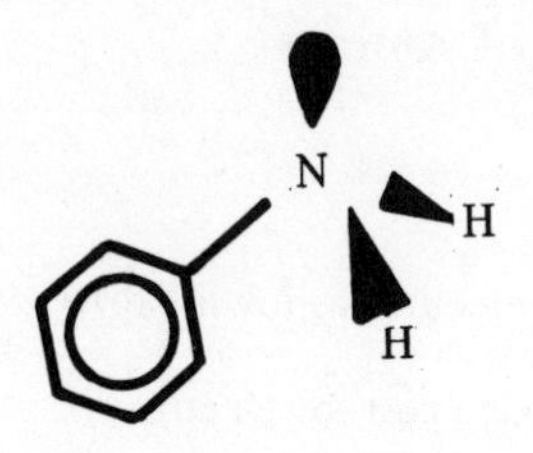

Chemical Properties of Amines

Comparison of ammonia and ethylamine (as an example of an aliphatic amine)

1. **Reaction with water.** Ammonia and aliphatic amines are very soluble in water forming alkaline solutions.

$$(a)\ NH_3 + H_2O \rightleftharpoons NH_4^+{}_{(aq)} + OH^-_{(aq)}$$

$$(b)\ C_2H_5NH_2 + H_2O \rightleftharpoons C_2H_5NH_3^+{}_{(aq)} + OH^-_{(aq)}$$

The vapours above solutions of these substances can be shown to be alkaline by testing with moist pH paper.

2. **Reaction with $CuSO_4$ solution.** Solutions of ammonia and amines react with copper(II) sulphate solution to form at first a light-blue precipitate of copper(II) hydroxide. This dissolves when excess ammonia or amine is added to form a very deep blue solution.

3. **Reaction with acids.** Solutions of ammonia and amines neutralise acids to form salts.

(a) $NH_{3\,(aq)} + H^+_{(aq)} + Cl^-_{(aq)} \longrightarrow NH^+_{4\,(aq)} + Cl^-_{(aq)}$

(b) $C_2H_5NH_{2\,(aq)} + H^+_{(aq)} + Cl^-_{(aq)} \longrightarrow C_2H_5NH^+_{3\,(aq)} + Cl^-_{(aq)}$

Careful evaporation of the solutions formed will yield samples of white crystalline salts, (a) ammonium chloride, NH_4Cl and (b) ethylammonium chloride, $C_2H_5NH_3Cl$.

4. **Effect of hot alkali on ammonium and amine salts.** When an ammonium salt is heated in the presence of a strong alkali, ammonia gas is evolved. Similarly, ethylamine can be liberated from an ethylammonium salt.

(a) $NH_4Cl + OH^-_{(aq)} \longrightarrow NH_{3\,(g)} + H_2O_{(l)} + Cl^-_{(aq)}$

(b) $C_2H_5NH_3Cl + OH^-_{(aq)} \longrightarrow C_2H_5NH_{2\,(g)} + H_2O_{(l)} + Cl^-_{(aq)}$

The reactions described above indicate that the structural resemblance between ammonia and amines is reflected in similar chemical behaviour. This is not too surprising when we consider that the reactions mentioned above involve directly the 'lone pair' of electrons on the nitrogen atom.

When ammonia or an amine dissolves in water, the 'lone pair' of electrons on the nitrogen atom extracts a proton from a water molecule and forms a co-ordinate bond with it, as shown in Figs. 17.1 and 17.2.

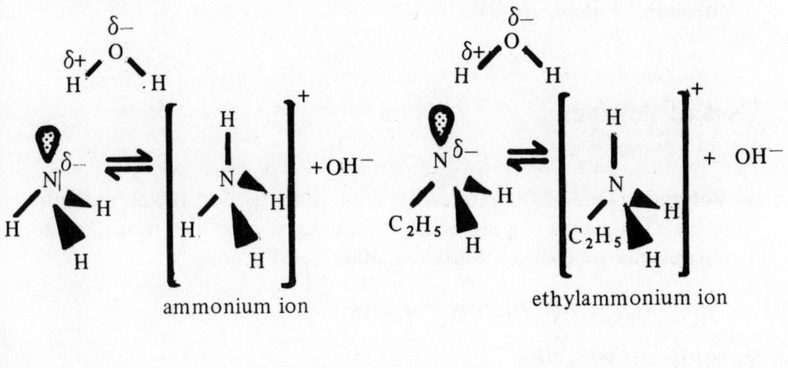

ammonium ion

ethylammonium ion

Fig. 17.1

These solutions are not fully dissociated and are, therefore, weak bases. If we compare solutions of equal concentration, they are found to be poorer conductors of electricity than strong alkalis such as NaOH and KOH.

Reactions 3 and 4 can be explained by reference to the equilibrium that exists between molecules and ions as illustrated in Figs. 17.1 and 17.2. The addition of acid causes the removal of hydroxyl ions and adjusts the equilibrium to the right. If sufficient acid is used the forward reaction will go to completion and the base is neutralised. When alkali is added to the ammonium or amine salts, the reverse reaction occurs. The action of heat helps to liberate the ammonia or amine in the form of a gas.

Aniline (Phenylamine)

Like most aromatic compounds, aniline is not soluble in water. It does, however, dissolve in acids to form salts. The solution formed no longer smells of aniline.

$$C_6H_5-NH_{2(l)} + H^+_{(aq)} + Cl^-_{(aq)} \longrightarrow C_6H_5-NH_3^+{}_{(aq)} + Cl^-_{(aq)}$$

If alkali is added to this solution, aniline is reformed as shown by the presence of oily droplets and aniline's characteristic smell.

$$C_6H_5-NH_3^+{}_{(aq)} + OH^-_{(aq)} \longrightarrow C_6H_5-NH_{2(l)} + H_2O_{(l)}$$

Although aniline does not bear such a strong resemblance to ammonia as do the simpler aliphatic amines, these reactions show that it does have basic properties, namely, (a) it accepts protons from acids to form cations which in turn (b) yield their extra protons to alkalis. As with ammonia and other amines, the 'lone pair' of electrons on the nitrogen atom is responsible for these effects.

Uses of Amines

1. Polymers: One of the monomers commonly used in the manufacture of nylon is a diamine, i.e. a molecule containing two amine groups, called 1,6-diaminohexane (hexamethylene diamine). Formula:

$H_2NCH_2CH_2CH_2CH_2CH_2CH_2NH_2$

or $H_2N(CH_2)_6NH_2$

Amines are also used to make the monomers required in the production of polyurethane.

2. Anti-oxidants: Amines are added to rubber to inhibit oxidation.

3. Medicine: A wide variety of amines is used in medicine, e.g. anti-malarial drugs and sulphanilamides.

4. Dyes: Many amines are used in the manufacture of dyes, e.g. 'azo dyes' derived from aromatic amines.

Proteins and Amino Acids

When proteins are hydrolysed, compounds known as **amino acids** are produced. Hydrolysis can be achieved either by using enzymes or by prolonged heating with acid. The amino acids can then be separated by chromatography. The general formula of an amino acid is shown below.

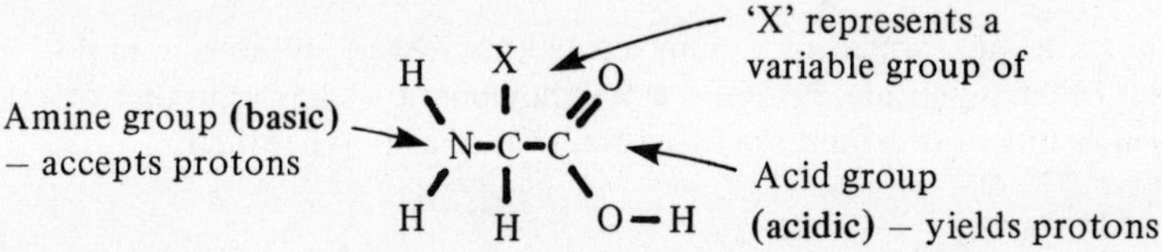

$$H_2N-CHX-C(=O)-O-H$$

Amine group **(basic)** – accepts protons

'X' represents a variable group of

Acid group **(acidic)** – yields protons

e.g. Glycine (X=H), the simplest amino acid, and alanine ($X=CH_3$).

$$H_2N-CH_2-C(=O)-O-H \qquad H_2N-CH(CH_3)-C(=O)-O-H$$

It is possible for proton transfer to occur within a single molecule. This gives rise to an ion with different charges at its two ends, e.g. with glycine –

$$NH_2-CH_2-COOH \longrightarrow \overset{+}{N}H_3-CH_2-COO^-$$

(proton transfer)

Amino acid molecules join together by a condensation reaction to form proteins. The reaction is similar to the formation of esters, except that, in this case, the condensation occurs at both ends of the molecule, as shown in Fig. 17.3.

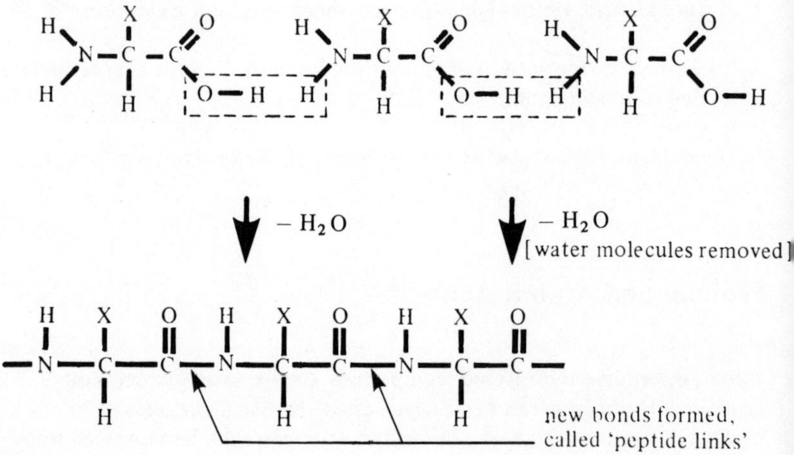

$- H_2O$ $- H_2O$ [water molecules removed]

new bonds formed, called 'peptide links'

Fig. 17.3

The molecule formed by joining a small number of amino acid molecules is called a **peptide.** Proteins, since they contain a large number of molecules linked together, are often referred to as **polypeptides.**

Importance of proteins

When digested, proteins are hydrolysed by enzymes to give amino acids, which are carried round the body in the blood stream and resynthesised into body protein such as muscle, hair, nails. Red blood cells contain the vitally important substance haemoglobin. This material consists of disc-shaped molecules, called 'haem', with iron ions at their centres, which act as 'oxygen carriers', attached to the 'globin' part which is a complex protein chain.

In recent years, methods have been developed for the synthesis of protein from hydrocarbons present in oil. The process involves the fermentation of hydrocarbons by yeast in the presence of air, water, ammonia and various mineral nutrients. The protein is then used to feed animals. At Grangemouth, a 4000 ton per annum plant, which was opened in 1971, manufactures protein from a feedstock consisting of straight-chain alkanes with between 10 and 23 carbon atoms per molecule. Larger industrial plants have been constructed in Southern France and Sardinia. It may well be that this process will play a major role in man's quest to solve the problem of the world shortage of protein.

Examples for practice

1.

$$\underset{A}{C_3H_6} \xrightarrow{1} \underset{B}{C_3H_7Br} \xrightarrow{2} \underset{C}{C_3H_7OH} \xrightarrow{3} \underset{D}{C_3H_6O}$$

The above reaction scheme outlines the conversion of an alkene A to a ketone D.
(i) Draw structural formulae for A, B, C, and D.
(ii) Name the type of change which takes place at 1 , 2 , and 3 respectively.
(iii) State the reagent and conditions for the reaction at 2.
(iv) Outline how C could be changed back to A.
(v) What, if anything, would be observed on addition of
(a) silver nitrate solution to B, (b) bromine water to A, (c) sodium to C?
(vi) When A is heated under pressure in the presence of a titanium catalyst it polymerises to form a useful plastic material E.
Name E, and make a diagram showing its structure. (At least three monomer units should be shown.) (S.C.E.E.B.)

2. (a) What is meant by the terms
(i) homologous series, (ii) isomerism?
(b) From the Heats of Combustion on page 202, choose values for three members of each of two homologous series. Use these figures to support the statement

> 'There is a regular increase in the heat of combustion as you move up an homologous series. The increase is the same for all such series.'

Give a name for each of the two series you choose.
(c) There are four isomeric alcohols of molecular formula C_4H_9OH. Their structural formulae are as follows:

$CH_3CH_2CH_2CH_2OH$ (I)

$CH_3CH_2\underset{|\atop OH}{C}HCH_3$ (II)

$CH_3-\overset{CH_3\atop |}{\underset{|\atop OH}{C}}-CH_3$ (III)

$CH_3\overset{CH_3\atop |}{C}HCH_2OH$ (IV)

(i) Give systematic names for (I), (II), (III), and (IV).
(ii) State which of the compounds I-IV are primary, which secondary, and which tertiary alcohols.

(iii) The four alcohols are contained, separately, in four bottles marked A, B, C, and D. From the following information decide which bottle contains which alcohol. State your reasoning briefly at each stage.

(*α*) The contents of A, B, and C can readily be oxidised by acid potassium dichromate solution, while those of D cannot.

(*β*) A and B, on complete oxidation by the dichromate, give acids of formulae C_3H_7COOH. C does not give this acid.

(*γ*) All four substances can be dehydrated to give alkenes. A and D can both form the same alkene.

B and C can both form the same alkene, which is an isomer of that formed by A and D. (S.C.E.E.B.)

3.

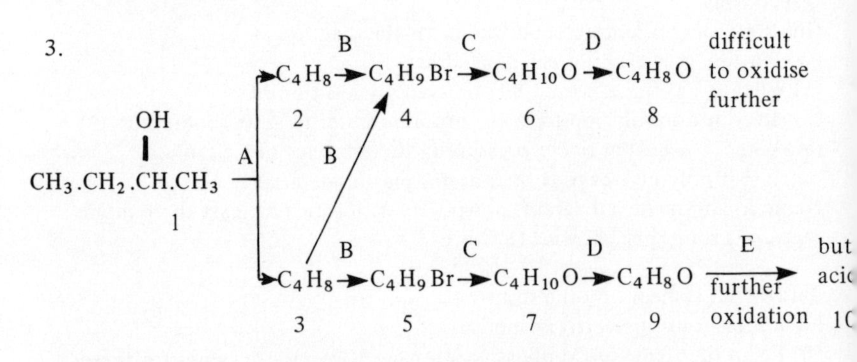

(a) Is compound 1 a primary, secondary or tertiary alcohol?

(b) Draw structural formulae for the isomeric compounds 2 and 3 .

(c) What reagents would be used in steps A, B, C and E?

(d) Draw structural formulae for 8 and 9 and name a reagent (other than E) which you could use to distinguish between them.

(e) Which of compounds 6 and 7 is an isomer of 1 ?

(f) Which compound 4 , 6 or 8 would react with sodium to produce hydrogen?

(g) Experiment shows that butanoic acid has a formula weight of approximately twice the expected value. Offer an explanation.

(h) Draw the structural formula for the ester formed by the action of 1 and 10 .

(i) When substance 1 is mixed with water, there is a slight increase in temperature, but when 1 is mixed with cyclohexane, C_6H_{12}, there is a slight decrease in temperature. Explain these observations with reference to the breaking and making of hydrogen bonds. (S.C.E.E.B.)

4. Parts (a), (b), (c) and (d) refer to the following reaction sequence.

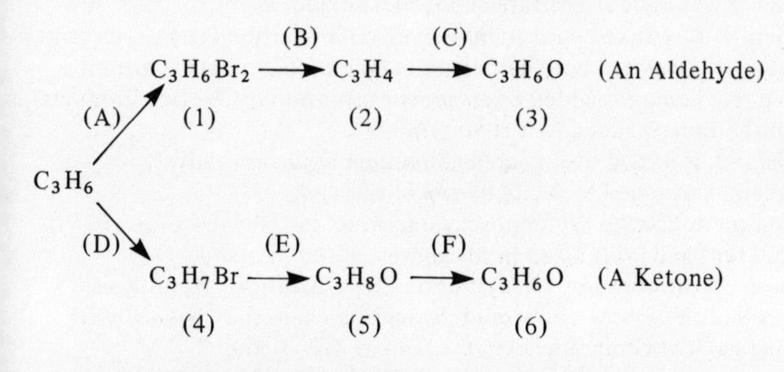

(a) Name compounds (3) and (5).
(b) Draw structural formulae of (3), (6), (1) and (4).
(c) What reagents would achieve steps (B), (D), (E) and (F)?
(d) Describe a test which would distinguish between the isomeric compounds (3) and (6).
(e) An ester can be made by the reaction of methanol with propanoic acid.
(i) Draw the structural formula of the ester and give its name.
(ii) State briefly how you would carry out the reaction. How would you ensure that it went to completion? (S.C.E.E.B.)

5.

1. $CH_3CH_2CH_2OH$
2. $CH_3CH_2CH(OH)CH_2CH_3$
3. $(CH_3CH_2)(CH_3)C{=}O$
4. CH_3NH_2
5. $CH_3C({=}O)OH$
6. $CH_3CH_2C({=}O)H$
7. $CH_3CH_2C({=}O)O{-}CH_3$
8. $CH_3CH_2C(CH_3)(OH)CH_2CH_3$

Answer the following questions by referring to the structural formulae shown above.
(a) Which of the above substances is (i) an aldehyde, (ii) an ester, (iii) an amine, (iv) a secondary alcohol, (v) a tertiary alcohol?
(b) Classify the remaining three compounds.
(c) Give the systematic name of each compound.
(d) Name and give the formula of the main product of each of the following reactions:-

(i) reaction between 1 and 5 ;
(ii) when 2 is passed over heated copper(II) oxide;
(iii) when 7 is refluxed with sodium hydroxide solution (two products);
(iv) when 6 is warmed with acidified potassium dichromate solution;
(v) when magnesium is added to an aqueous solution of 5 (two products);
(vi) when sodium is added to 1 (two products);
(vii) when 2 is passed over heated aluminium oxide as catalyst;
(viii) when 4 is added to dilute hydrochloric acid.
(e) From the following list, choose one term to describe reactions (i), (ii), (iii), (iv), (vii) and (viii) listed in (d) above.
Oxidation ; Neutralisation ; Dehydration ; Esterification ; Hydrolysis.
(f) Describe briefly how you would distinguish chemically between the following pairs of compounds:- (i) 3 and 6 (ii) 5 and 7
(iii) 2 and 8 (iv) 1 and 6 (v) 4 and 5 .
(g) Which of the above compounds could give on reduction
(i) ethanal (ii) propan-l-ol (iii) butan-2-ol?
(h) Give the structural formula and systematic name of an isomer of
(i) 2 and (ii) 8 .

6. Describe how you could separate a mixture of toluene, phenol and aniline so that you obtain a sample of each. The samples need not be purified. Assume that you have available dilute hydrochloric acid, sodium hydroxide solution and a separating funnel in addition to normal laboratory apparatus.

Data Tables

Periodic Table

Col. 1 Col. 2

Symbol	Atomic Number / Atomic Weight
	Electron arrangement
	Density (kg m^{-3})*
	Name

Col. 1 and Col. 2

Symbol	Atomic Number	Atomic Weight	Electron arrangement	Density (kg m^{-3})*	Name
H	1	1·01	1	0·0899	Hydrogen
Li	3	6·94	2, 1	$0·53 \times 10^3$	Lithium
Be	4	9·01	2, 2	$1·85 \times 10^3$	Beryllium
Na	11	23·0	2, 8, 1	$0·97 \times 10^3$	Sodium
Mg	12	24·3	2, 8, 2	$1·74 \times 10^3$	Magnesium
K	19	39·1	2, 8, 8, 1	$0·86 \times 10^3$	Potassium
Ca	20	40·1	2, 8, 8, 2	$1·54 \times 10^3$	Calcium
Rb	37	85·5	2, 8, 18, 8, 1	$1·53 \times 10^3$	Rubidium
Sr	38	87·6	2, 8, 18, 8, 2	$2·60 \times 10^3$	Strontium
Cs	55	133	2, 8, 18, 18, 8, 1	$1·88 \times 10^3$	Caesium
Ba	56	137	2, 8, 18, 18, 8, 2	$3·51 \times 10^3$	Barium
Fr	87		2, 8, 18, 32, 18, 8, 1		Francium
Ra	88	226	2, 8, 18, 32, 18, 8, 2	$5·00 \times 10^3$	Radium

TRANSITION ELEMENTS

Symbol	Atomic Number	Atomic Weight	Electron arrangement	Density (kg m^{-3})*	Name
Sc	21	45·0	2, 8, 9, 2	$2·99 \times 10^3$	Scandium
Ti	22	47·9	2, 8, 10, 2	$4·50 \times 10^3$	Titanium
V	23	50·9	2, 8, 11, 2	$5·96 \times 10^3$	Vanadium
Cr	24	52·0	2, 8, 13, 1	$7·20 \times 10^3$	Chromium
Mn	25	54·9	2, 8, 13, 2	$7·20 \times 10^3$	Manganese
Fe	26	55·8	2, 8, 14, 2	$7·86 \times 10^3$	Iron
Co	27	58·9	2, 8, 15, 2	$8·90 \times 10^3$	Cobalt
Y	39	88·9	2, 8, 18, 9, 2	$4·47 \times 10^3$	Yttrium
Zr	40	91·2	2, 8, 18, 10, 2	$7·09 \times 10^3$	Zirconium
Nb	41	92·9	2, 8, 18, 11, 2	$8·57 \times 10^3$	Niobium
Mo	42	95·9	2, 8, 18, 13, 1	$10·2 \times 10^3$	Molybdenum
Tc	43	98·9	2, 8, 18, 14, 1	$11·5 \times 10^3$	Technetium
Ru	44	101	2, 8, 18, 15, 1	$12·3 \times 10^3$	Ruthenium
Rh	45	103	2, 8, 18, 16, 1	$12·4 \times 10^3$	Rhodium
La	57	139	2, 8, 18, 18, 9, 2	$6·19 \times 10^3$	Lanthanum
Hf	72	178	2, 8, 18, 32, 10, 2	$13·3 \times 10^3$	Hafnium
Ta	73	181	2, 8, 18, 32, 11, 2	$16·6 \times 10^3$	Tantalum
W	74	184	2, 8, 18, 32, 12, 2	$19·4 \times 10^3$	Tungsten
Re	75	186	2, 8, 18, 32, 13, 2	$20·5 \times 10^3$	Rhenium
Os	76	190	2, 8, 18, 32, 14, 2	$22·5 \times 10^3$	Osmium
Ir	77	192	2, 8, 18, 32, 17, 0	$22·4 \times 10^3$	Iridium
Ac	89		2, 8, 18, 32, 18, 9, 2		Actinium

LANTHANIDES:

Symbol	Atomic Number	Atomic Weight	Electron arrangement	Name
La	57	139	2, 8, 18, 18, 9, 2	Lanthanum
Ce	58	140	2, 8, 18, 20, 8, 2	Cerium
Pr	59	140	2, 8, 18, 21, 8, 2	Praseodymium
Nd	60	144	2, 8, 18, 22, 8, 2	Neodymium
Pm	61		2, 8, 18, 23, 8, 2	Promethium
Sm	62	150	2, 8, 18, 24, 8, 2	Samarium

ACTINIDES:

Symbol	Atomic Number	Atomic Weight	Electron arrangement	Name
Ac	89		2, 8, 18, 32, 18, 9, 2	Actinium
Th	90	232	2, 8, 18, 32, 18, 10, 2	Thorium
Pa	91	231	2, 8, 18, 32, 20, 9, 2	Protactinium
U	92	238	2, 8, 18, 32, 21, 9, 2	Uranium
Np	93	237	2, 8, 18, 32, 22, 9, 2	Neptunium
Pu	94		2, 8, 18, 32, 24, 8, 2	Plutonium

*Gases at standard temperature and pressure

			Col. 3	Col. 4	Col. 5	Col. 6	Col. 7	Col. 8
								He 2 4·00 2 0·179 Helium
			B 5 10·8 2, 3 2·34 × 10³ Boron	C 6 12·0 2, 4 2·25 × 10³ Carbon	N 7 14·0 2, 5 1·25 Nitrogen	O 8 16·0 2, 6 1·43 Oxygen	F 9 19·0 2, 7 1·70 Fluorine	Ne 10 20·2 2, 8 0·90 Neon
			Al 13 27·0 2, 8, 3 2·70 × 10³ Aluminium	Si 14 28·1 2, 8, 4 2·33 × 10³ Silicon	P 15 31·0 2, 8, 5 1·82 × 10³ Phosphorus	S 16 32·1 2, 8, 6 2·07 × 10³ Sulphur	Cl 17 35·5 2, 8, 7 3·21 Chlorine	Ar 18 39·9 2, 8, 8 1·78 Argon
Ni 28 58·7 2, 8, 16, 2 8·90 × 10³ Nickel	Cu 29 63·5 2, 8, 18, 1 8·92 × 10³ Copper	Zn 30 65·4 2, 8, 18, 2 7·14 × 10³ Zinc	Ga 31 69·7 2, 8, 18, 3 5·90 × 10³ Gallium	Ge 32 72·6 2, 8, 18, 4 5·35 × 10³ Germanium	As 33 74·9 2, 8, 18, 5 5·73 × 10³ Arsenic	Se 34 79·0 2, 8, 18, 6 4·81 × 10³ Selenium	Br 35 80·0 2, 8, 18, 7 3·12 × 10³ Bromine	Kr 36 83·8 2, 8, 18, 8 3·71 Krypton
Pd 46 106 2, 8, 18, 18, 0 12·0 × 10³ Palladium	Ag 47 108 2, 8, 18, 18, 1 10·5 × 10³ Silver	Cd 48 112 2, 8, 18, 18, 2 8·64 × 10³ Cadmium	In 49 115 2, 8, 18, 18, 3 7·31 × 10³ Indium	Sn 50 119 2, 8, 18, 18, 4 7·28 × 10³ Tin	Sb 51 122 2, 8, 18, 18, 5 6·68 × 10³ Antimony	Te 52 128 2, 8, 18, 18, 6 6·25 × 10³ Tellurium	I 53 127 2, 8, 18, 18, 7 4·93 × 10³ Iodine	Xe 54 131 2, 8, 18, 18, 8 5·85 Xenon
Pt 78 195 2, 8, 18, 32, 17, 1 21·5 × 10³ Platinum	Au 79 197 2, 8, 18, 32, 18, 1 19·3 × 10³ Gold	Hg 80 201 2, 8, 18, 32, 18, 2 13·6 × 10³ Mercury	Tl 81 204 2, 8, 18, 32, 18, 3 11·8 × 10³ Thallium	Pb 82 207 2, 8, 18, 32, 18, 4 11·3 × 10³ Lead	Bi 83 209 2, 8, 18, 32, 18, 5 9·80 × 10³ Bismuth	Po 84 2, 8, 18, 32, 18, 6 9·4 × 10³ Polonium	At 85 2, 8, 18, 32, 18, 7 Astatine	Rn 86 2, 8, 18, 32, 18, 8 Radon

Eu 63 152 2, 8, 18, 25, 8, 2 Europium	Gd 64 157 2, 8, 18, 25, 9, 2 Gadolinium	Tb 65 159 2, 8, 18, 27, 8, 2 Terbium	Dy 66 162 2, 8, 18, 28, 8, 2 Dysprosium	Ho 67 165 2, 8, 18, 29, 8, 2 Holmium	Er 68 167 2, 8, 18, 30, 8, 2 Erbium	Tm 69 169 2, 8, 18, 31, 8, 2 Thulium	Yb 70 173 2, 8, 18, 32, 8, 2 Ytterbium	Lu 71 175 2, 8, 18, 32, 9, 2 Lutetium
Am 95 2, 8, 18, 32, 25, 8, 2 Americium	Cm 96 2, 8, 18, 32, 25, 9, 2 Curium	Bk 97 2, 8, 18, 32, 26, 9, 2 Berkelium	Cf 98 2, 8, 18, 32, 28, 8, 2 Californium	Es 99 2, 8, 18, 32, 29, 8, 2 Einsteinium	Fm 100 2, 8, 18, 32, 30, 8, 2 Fermium	Md 101 2, 8, 18, 32, 31, 8, 2 Mendelevium	No 102 2, 8, 18, 32, 32, 8, 2 Nobelium	Lr 103 2, 8, 18, 32, 32, 9, 2 Lawrencium

Properties of the Elements

Symbol	Atomic Number
	m.p. (°C) b.p. (°C)
1st, 2nd, 3rd, 4th Ionisation Energies ($J\ mol^{-1} \times 10^3$)	Electro-negativity (Pauling) Covalent Radius ($m \times 10^{-10}$) [a] (Charge on Ion) Ionic Radius [b] ($m \times 10^{-10}$)

Symbol	Atomic Number	m.p. (°C)	b.p. (°C)	1st Ionisation	2nd Ionisation	3rd Ionisation	4th Ionisation	Electro-negativity	Covalent Radius	Charge on Ion	Ionic Radius
H	1	−259	−253	1310				2·1	0·37	(1−)	2·08
Li	3	179	1317	520	7280	11800		1·0	1·23	(1+)	0·68
Be	4	1280	2970	900	1760	14900		1·5	0·89	(2+)	0·35
Na	11	98	892	490	4560			0·9	1·57	(1+)	1·10
Mg	12	651	1110	740	1450	7740		1·2	1·36	(2+)	0·80
K	19	64	774	420	3070			0·8	2·03	(1+)	1·46
Ca	20	843	1490	590	1150	4940		1·0	1·74	(2+)	1·08
Sc	21	1540	2730	630	1240	2390	7120	1·3	1·44	(3+)	0·81
Ti	22	1680	3260	660	1310	2720	4170	1·5	1·32	(2+)	0·94
V	23	1890	3000	650	1370	2870	4600	1·6	1·22	(2+)	0·87
Cr	24	1890	2480	650	1590	2990	4770	1·6	1·17	(2+)	0·90
Mn	25	1240	2100	720	1510	3250	5190	1·5	1·17	(2+)	0·90
Fe	26	1540	3000	760	1560	2960	5400	1·8	1·16	(2+)	0·85
Co	27	1500	2900	760	1650	3230	5100	1·8	1·16	(2+)	0·82
Rb	37	39	688	400	2650			0·8	2·16	(1+)	1·57
Sr	38	769	1380	550	1060			1·0	1·91	(2+)	1·24
Y	39	1500	2930	640	1180			1·2	1·62	(3+)	0·97
Zr	40	2980		670	1270			1·4	1·45		
Nb	41	2470	4930	650	1380			1·6	1·34		
Mo	42	2610	5560	690	1560			1·8	1·29		
Tc	43	2200	3500	700	1470			1·9			
Ru	44	2250	3900	720	1620			2·2	1·24		
Rh	45	1970	3730	750	1750			2·2	1·25		
Cs	55	29	690	380	2420			0·7	2·35	(1+)	1·78
Ba	56	725	1140	500	970			0·9	1·98	(2+)	1·44
La	57	920	3470	540	1100			1·1	1·69	(3+)	1·14
Hf	72	2150	5400	530	1440			1·3			
Ta	73	2300	5430	580	1570			1·5	1·44		
W	74	3410	5930	770	1710			1·7	1·30		
Re	75	3180	5630	760	1600			1·9	1·28		
Os	76	3000	5000	840	1630			2·2	1·26		
Ir	77	2410	4530	890				2·2	1·26		
Fr	87							0·7			
Ra	88	700	1750	510	980			0·9			
Ac	89	3200		670	1170			1·1			

Notes

[a] The Ionic Charge quoted for an element in this table should not be taken to mean that the element always exists as that ion in its compounds, e.g.

hydrogen does not usually form H^-;

carbon does not usually form C^{4+}.

[b] The ionic radii presented are "rationalised" values, some of which may differ from values given in certain textbooks.

								He 2 −272 −269 2370 5250
			B 5 2300 2550 800 2·0 2430 0·80 3660 (3+) 25000 0·20	**C** 6 3650 4200 1090 2·5 2350 0·77 4600 (4+) 6240 0·15	**N** 7 −210 −196 1400 3·0 2860 0·74 4600 (3−) 7480 1·71	**O** 8 −218 −183 1310 3·5 3390 0·74 5320 (2−) 7460 1·30	**F** 9 −220 −188 1680 4·0 3370 0·72 6030 (1−) 8410 1·23	**Ne** 10 −249 −246 2080 3960 6200 9380
			Al 13 660 2470 580 1·5 1820 1·25 2750 (3+) 11600 0·61	**Si** 14 1410 2360 790 1·8 1580 1·17 3230 (4+) 0·48	**P** 15 44 280 1060 2·1 1900 1·10 2920 (3−) 2·12	**S** 16 113 445 970 2·5 2260 1·04 3390 (2−) 1·72	**Cl** 17 −101 −35 1260 3·0 2300 0·99 3850 (1−) 1·72	**Ar** 18 −189 −186 1520
Ni 28 1450 2730 740 1·8 1750 1·15 3390 (2+) 5400 0·78	**Cu** 29 1080 2600 750 1·9 1960 1·17 3550 (2+) 5700 0·81	**Zn** 30 419 907 910 1·6 1730 1·25 3830 (2+) 5990 0·83	**Ga** 31 30 2400 580 1·6 1980 1·25 2960 (3+) 6200 0·70	**Ge** 32 937 2830 760 1·8 1540 1·22 3300 (4+) 0·62	**As** 33 817 613 970 2·0 1950 1·21 2730 (3−) 2·22	**Se** 34 217 685 940 2·4 2080 1·17 3090 (2−) 1·88	**Br** 35 −7 59 1140 2·8 2080 1·14 3470 (1−) 1·88	**Kr** 36 −157 −152 1350
Pd 46 1550 2930 800 2·2 1880 1·28 (2+) 0·94	**Ag** 47 960 2210 730 1·9 2070 1·34 (1+) 1·23	**Cd** 48 321 765 870 1·7 1630 1·41 (2+) 1·03	**In** 49 157 2000 560 1·7 1820 1·50 (3+) 0·87	**Sn** 50 232 2260 710 1·8 1410 1·40 2940 (2+) 3930 1·12	**Sb** 51 630 1380 830 1·9 1590 1·41 (3−) 2·45	**Te** 52 450 990 870 2·1 1800 1·37 (2−) 2·21	**I** 53 114 184 1010 2·5 1840 1·33 (1−) 2·13	**Xe** 54 −112 −107 1170
Pt 78 1770 3900 870 2·2 1870 1·29	**Au** 79 1060 2970 890 2·4 1980 1·34	**Hg** 80 −39 357 1010 1·9 1810 1·44 (2+) 1·10	**Tl** 81 304 1460 590 1·8 1970 1·55 (3+) 0·96	**Pb** 82 328 1740 720 1·8 1450 1·54 (2+) 1·26	**Bi** 83 271 1560 770 1·9 1610 1·52	**Po** 84 254 962 810 2·0	**At** 85 2·2 1·40	**Rn** 86 −71 −62 1040

Heats of Combustion

C(s)	−394 kJ mol^{-1}	C_6H_6(l) (benzene)	−3270 kJ mol^{-1}
		C_6H_{12}(l) (cyclohexane)	−3920
H_2(g)	−286	C_6H_{10}(l) (cyclohexene)	−3730
CH_4(g)	−890		
C_2H_6(g)	−1560	CH_3OH(l)	−715
C_3H_8(g)	−2220	C_2H_5OH(l)	−1370
C_4H_{10}(g)	−2880	C_3H_7OH(l)	−2010
		C_4H_9OH(l)	−2670
C_8H_{18}(l)	−5510		
C_2H_4(g)	−1410		
C_2H_2(g)	−1300		

Mean Bond Dissociation Energies

C − H	414 kJ mol^{-1}	F − F	155 kJ mol^{-1}
C − C (aliphatic)	347	Cl − Cl	243
C ∵ C (aromatic)	519	Br − Br	192
C = C	598	I − I	151
C ≡ C	811		
		H − O	465
C − F	486	C − O	360
C − Cl	339	C = O	744
C − Br	277	O = O	495
C − I	239		
H − H	435		
H − F	565		
H − Cl	431		
H − Br	364		
H − I	297		

Heat of Sublimation of Carbon

This is the energy required to form separate atoms of carbon from solid carbon

$$C(s) \longrightarrow C(g) \qquad \Delta H = 715 \text{ kJ mol}^{-1}$$

Standard Reduction Electrode Potentials (I.U.P.A.C. convention)

Reaction		E°
$Li^{+}(aq) + e$	$\rightarrow Li(s)$	−3·02 V
$Cs^{+}(aq) + e$	$\rightarrow Cs(s)$	−2·92
$Rb^{+}(aq) + e$	$\rightarrow Rb(s)$	−2·92
$K^{+}(aq) + e$	$\rightarrow K(s)$	−2·92
$Ca^{2+}(aq) + 2e$	$\rightarrow Ca(s)$	−2·87
$Na^{+}(aq) + e$	$\rightarrow Na(s)$	−2·71
$Mg^{2+}(aq) + 2e$	$\rightarrow Mg(s)$	−2·37
$Al^{3+}(aq) + 3e$	$\rightarrow Al(s)$	−1·66
$Zn^{2+}(aq) + 2e$	$\rightarrow Zn(s)$	−0·76
$Cr^{3+}(aq) + 3e$	$\rightarrow Cr(s)$	−0·74
$S(s) + 2e$	$\rightarrow S^{2-}(aq)$	−0·51
$Fe^{2+}(aq) + 2e$	$\rightarrow Fe(s)$	−0·44
$Cr^{3+}(aq) + e$	$\rightarrow Cr^{2+}(aq)$	−0·41
$Sn^{2+}(aq) + 2e$	$\rightarrow Sn(s)$	−0·14
$Pb^{2+}(aq) + 2e$	$\rightarrow Pb(s)$	−0·13
$Fe^{3+}(aq) + 3e$	$\rightarrow Fe(s)$	−0·04
$2H^{+}(aq) + 2e$	$\rightarrow H_2(g)$	0·00
$S(s) + 2H^{+}(aq) + 2e$	$\rightarrow H_2S(aq)$	0·14
$Sn^{4+}(aq) + 2e$	$\rightarrow Sn^{2+}(aq)$	0·15
$Cu^{2+}(aq) + e$	$\rightarrow Cu^{+}(aq)$	0·15
$SO_4{}^{2-}(aq) + 2H^{+}(aq) + 2e$	$\rightarrow SO_3{}^{2-}(aq) + H_2O$	0·17
$Hg_2Cl_2(s) + 2e$	$\rightarrow 2Hg(l) + 2Cl^{-}(aq)$	0·27
$Cu^{2+}(aq) + 2e$	$\rightarrow Cu(s)$	0·34
$Cu^{+}(aq) + e$	$\rightarrow Cu(s)$	0·52
$I_2(s) + 2e$	$\rightarrow 2I^{-}(aq)$	0·54
$Fe^{3+}(aq) + e$	$\rightarrow Fe^{2+}(aq)$	0·77
$Ag^{+}(aq) + e$	$\rightarrow Ag(s)$	0·80
$2NO_3{}^{-}(aq) + 4H^{+}(aq) + 2e$	$\rightarrow N_2O_4(g) + 2H_2O$	0·81
$Hg^{2+}(aq) + 2e$	$\rightarrow Hg(l)$	0·85
$NO_3{}^{-}(aq) + 4H^{+}(aq) + 3e$	$\rightarrow NO(g) + 2H_2O$	0·96
$Br_2(l) + 2e$	$\rightarrow 2Br^{-}(aq)$	1·07
$O_2(g) + 4H^{+}(aq) + 4e$	$\rightarrow 2H_2O$	1·23
$MnO_2(s) + 4H^{+}(aq) + 2e$	$\rightarrow Mn^{2+}(aq) + 2H_2O$	1·23
$Cr_2O_7{}^{2-}(aq) + 14H^{+}(aq) + 6e$	$\rightarrow 2Cr^{3+}(aq) + 7H_2O$	1·33
$Cl_2(aq) + 2e$	$\rightarrow 2Cl^{-}(aq)$	1·36
$Au^{3+}(aq) + 3e$	$\rightarrow Au(s)$	1·50
$MnO_4{}^{-}(aq) + 8H^{+}(aq) + 5e$	$\rightarrow Mn^{2+}(aq) + 4H_2O$	1·51
$Au^{+}(aq) + e$	$\rightarrow Au(s)$	1·68
$F_2(g) + 2e$	$\rightarrow 2F^{-}(aq)$	2·87

Note

These potentials refer to standard state conditions and this is of particular importance when the data are applied to ionic solutions.

Answers to Numerical Questions

Chapter 1

1. (a) Fr^+ (b) $^{223}_{88}Ra$
2. (a) $^{231}_{91}Pa$ (b) $^{227}_{89}Ac$ (c) α or $^{4}_{2}He$ (d) β or $^{0}_{-1}e$
3. (a) $^{234}_{91}Pa$ (b) $^{218}_{84}Po$
4. x : $^{4}_{2}He$ y : $^{230}_{90}Th$ z : $^{212}_{82}Pb$
5. 8 days
6. 1/64

Chapter 2

1. 65.46 2. 107.97 3. 207.24

Chapter 3

1. (a) 3×10^{23} (b) 3×10^{23} (c) 2×10^{23}
2. (a) 1.5×10^{23} (b) 1.5×10^{22} (c) 3×10^{24}
   (d) 1.2×10^{23} (e) 3×10^{23}
3. (a) 1.8×10^{23} (b) 3.6×10^{23} (c) 7.5×10^{23}
4. 3×10^{25}
5. (a) 73 g (b) 254 g (c) 87 g (d) 250 g
6. (a) 76.5 g (b) 70 g (c) 6.3 g (d) 200 g
7. (a) 0.125 (b) 10 (c) 0.25 (d) 2
8. (a) 0.25 (b) 4 (c) 1 (d) 1.5
9. (a) C 52.2% H 13.0% O 34.8%
   (b) Fe 23.1% N 17.4% O 59.5%
   (c) Cu 40% S 20% O 40%
   (d) H 3.1% P 31.6% O 65.3%

10. urea 46.7% N ; ammonium nitrate 35% N.

11. (a) H_2SO_4 (b) C_6H_6O (c) $MgSO_4.7H_2O$ (d) $Pb(NO_3)_2$

12. C_2H_4

13. $KClO_3$

14. 5.55 g

15. 4.46 g

16. 13 g

17. 2.5 g

18. (a) 2.5 (b) 0.02 (c) 0.8 (d) 0.2

19. (a) 19.6 g (b) 2.38 kg (c) 10.2 g (d) 21.2 g

20. (a) 0.25 M (b) 1.33 M (c) 0.05 M (d) 0.5 M

21. (a) 40 cm^3 (b) 10 cm^3 (c) 20 cm^3

22. 0.4 M

23. (a) 2.32 g $litre^{-1}$ (b) 0.76 g $litre^{-1}$ (c) 0.89 g $litre^{-1}$

24. 28

25. (a) 5.6 litres (b) 4.48 litres (c) 2.24 litres

26. 9 g

27. 1.92 g

28. (a) 23.9 g (b) 2.24 litres (c) 0.2 M

29. (a) 2 litres O_2, 1 litre CO_2

(b) 3 litres O_2, 2 litres CO_2

(c) 1 litre O_2, 1 litre CO_2

30. 4.03 g

31. 8 g

32. 104 cm^3 H_2, 104 cm^3 Cl_2

33. 1.12 g

Chapter 4

1. $\Delta H = -1349$ kJ mol^{-1}

2. $\Delta H = -880$ kJ mol^{-1}

3. (a) $\Delta H_f = -251$ kJ mol^{-1} (b) $\Delta H_f = -160$ kJ mol^{-1}
4. $\Delta H = -849$ kJ mol^{-1}
5. (i) 364 kJ mol^{-1} (ii) $\Delta H_f = -50.5$ kJ mol^{-1}
6. $\Delta H = -606$ kJ mol^{-1}
7. (a) 327 kJ mol^{-1} (b) $\Delta H_f = -107$ kJ mol^{-1}
8. $\Delta H_f = -50$ kJ mol^{-1}
9. (a) $\Delta H = +243$ kJ mol^{-1} (b) $\Delta H = +4$ kJ mol^{-1}
   (c) $\Delta H = -188$ kJ mol^{-1}
10. $\Delta H = -322$ kJ mol^{-1}
11. (b) 304 kJ mol^{-1}
   (d) $\Delta H = -2095$ kJ mol^{-1}

Chapter 5

3. (a) 0.05 V (b) 0.94 V
6. (a) (iv) 0.44 V
7. (b) 0.67V

Chapter 15

4. (a) & (b) C_3H_8O
5. (e) 0.6 moles O_2

Index